# Managing Climate Change

# Managing Climate Change

Edited by Dale Sullivan

STATES
ACADEMIC PRESS
www.statesacademicpress.com

**States Academic Press,**
109 South 5th Street,
Brooklyn, NY 11249, USA

Visit us on the World Wide Web at:
www.statesacademicpress.com



ISBN: 978-1-63989-689-9



**Cataloging-in-publication Data**

Managing climate change / edited by Dale Sullivan.
p. cm.
Includes bibliographical references and index.
ISBN 978-1-63989-689-9
1. Climatic changes. 2. Climate change mitigation. 3. Climatology. I. Sullivan, Dale.
QC903 .C55 2023
577.22--dc23

# TABLE OF CONTENTS

**Preface** ... VII

Chapter 1 **Addressing Climate Change Vulnerability through Small Livestock Rearing in Matobo** ... 1
Keith Phiri, Sibonokuhle Ndlovu, Moreblessings Mpofu, Philani Moyo and Henri-Count Evans

Chapter 2 **Biochar for Climate Change Adaptation: Effect on Heavy Metal Composition of *Telfairia occidentalis* Leaves** ... 21
Doris Akachukwu, Michael Adedapo Gbadegesin, Philippa Chinyere Ojimelukwe and Christopher John Atkinson

Chapter 3 **Adaptation to Climate Change: Opportunities and Challenges from Zambia** ... 42
Jonty Rawlins and Felix Kanungwe Kalaba

Chapter 4 **Climate Change Adaptation: Opportunities for Increased Material Recycling Facilities in African Cities** ... 62
Gamuchirai Mutezo, Jean Mulopo and Dumisani Chirambo

Chapter 5 **African Yam Bean the Choice for Climate Change Resilience: Need for Conservation and Policy** ... 87
C. V. Nnamani, D. B. Adewale, H. O. Oselebe and C. J. Atkinson

Chapter 6 **Smallholders use of Weather Information as Smart Adaptation Strategy in the Savannah Area of Ondo State** ... 104
Rasheedat Alliagbor, David Olufemi Awolala and Igbekele Amos Ajibefun

Chapter 7 **Panel Analysis of the Relationship Between Weather Variability and Sectoral Output** ... 115
Olga Nekesa Mulama and Caroline Wanjiru Kariuki

Chapter 8 **Assessment of Farmers' Indigenous Technology Adoptions for Climate Change Adaptation** ... 131
Idowu Ologeh, Francis Adesina and Victor Sobanke

Chapter 9 **Climate Change, Rural Livelihoods and Ecosystem Nexus: Forest Communities in Agroecological Zones** ... 144
Olushola Fadairo, Samuel Olajuyigbe, Tolulope Osayomi, Olufolake Adelakun, Olanrewaju Olaniyan, Siji Olutegbe and Oluwaseun Adeleke

Chapter 10 **Social Vulnerability of Rural Dwellers to Climate Variability: Akwa Ibom State**........167
Jemimah Timothy Ekanem and Idongesit Michael Umoh

Chapter 11 **Impacts of Climate Change to Poultry Production in Africa: Adaptation Options for Broiler Chickens**........190
M. O. Abioja and J. A. Abiona

Chapter 12 **Livestock Breeders' Adaptation to Climate Variability and Change in Morocco's Arid Rangelands**........212
Wadii Snaibi and Abdelhamid Mezrhab

**Permissions**

**List of Contributors**

**Index**

# Preface

The main aim of this book is to educate learners and enhance their research focus by presenting diverse topics covering this vast field. This is an advanced book which compiles significant studies by distinguished experts in the area of analysis. This book addresses successive solutions to the challenges arising in the area of application, along with it; the book provides scope for future developments.

Climate change refers to the long-term changes in weather patterns and temperature. These changes can be caused by natural processes, like variations in the solar cycle. The main cause of climate change is the burning of fossil fuels such as coal, natural gas and oil. Some of the prominent environmental impacts of climate change are the expansion of deserts and an increase in the frequency of wildfires. Climate change threatens people with water and food shortage, extreme heat, economic loss, and flooding. There are two approaches to strategically manage the hazards associated with climate change. These include the mitigation of greenhouse gases, which decreases the severity and frequency of climate hazards; and adaptation to climate change, which helps in decreasing the vulnerability to a specific climate hazard or hazards. This book contains some path-breaking studies on climate change. From theories to research to practical applications, studies related to all contemporary topics of relevance to the management of climate change have been included herein. The book will serve as a valuable source of reference for graduate and postgraduate students.

It was a great honour to edit this book, though there were challenges, as it involved a lot of communication and networking between me and the editorial team. However, the end result was this all-inclusive book covering diverse themes in the field.

Finally, it is important to acknowledge the efforts of the contributors for their excellent chapters, through which a wide variety of issues have been addressed. I would also like to thank my colleagues for their valuable feedback during the making of this book.

**Editor**

# 1

# Addressing Climate Change Vulnerability through Small Livestock Rearing in Matobo, Zimbabwe

Keith Phiri, Sibonokuhle Ndlovu, Moreblessings Mpofu,
Philani Moyo and Henri-Count Evans

## Contents

Introduction .
The Impact of Climate Change in Zimbabwe
 Climate Change and Vulnerability
 Adaptation Strategies in Zimbabwe
Study Site and Context
 Materials and Methods
Results and Discussions
 Small Livestock and Climate Change Nexus in Matobo
Constraints to Small Livestock Rearing
Institutional Mandate Response on Small Livestock Production
Conclusions ....
Recommendations
References

K. Phiri (✉) · P. Moyo
Fort Hare Institute of Social and Economic Research (FHISER), University of Fort Hare, East London, South Africa
e-mail: kephirih@gmail.com; pmoyo@ufh.ac.za

S. Ndlovu · M. Mpofu
Department of Development Studies, Faculty of Humanities and Social Sciences, Lupane State University, Bulawayo, Zimbabwe
e-mail: boomagwala@gmail.com; moreblessingsncube@gmail.com

H.-C. Evans
School of Applied Human Sciences, Centre for Communication, Media and Society, University of Kwazulu-Natal, Durban, South Africa
e-mail: henricount@gmail.com

**Abstract**

Livestock rearing is a popular climate change adaptation strategy among farmers in Matobo District, Zimbabwe. In this chapter we reveal how farmers in Matobo District have benefited immensely from rearing small livestock in response to climate change. Although the descriptor "small livestock" generically refers to different types of livestock, in this chapter we limit it to goat and sheep rearing. The purpose of the chapter is (1) to discuss the efficacy of small livestock rearing as a response to climate change and (2) to use smallholder farmer's narratives to evaluate the success of government interventions in enhancing small livestock production. Utilizing an inductive approach, data was gathered through five (5) key informants, five (5) focus group discussions, and 50 in-depth semi-structured interviews. Our study reveals that small livestock are suitable and adaptable to climate change impacts in Matobo. We recommend that government and its development partners prioritize and avail funds for the increased uptake of small livestock rearing among smallholder farmers in Matobo District and beyond.

**Keywords**

Adaptation · Climate change · Small livestock · Smallholder farmers · Matobo · Zimbabwe

## Introduction

Small-scale communal livestock rearing plays an essential role in sustaining livelihoods in developing countries. Livestock is central to food security, nutrition, organic fertilizer (manure) production, non-mechanized ploughing, and income generation initiatives (Nandhini and Suganthi 2018). However, as climate change impacts are increasingly being felt across the world, poor countries are most affected owing to poor adaptive capacity and budgetary constraints compared to more developed countries in the global North (Chaudhury 2017). Developing countries' vulnerability is worsened by their strong dependence on rain-fed agriculture and free range pasture-fed livestock, which are susceptible to the negative effects of changing climate (Dube and Phiri 2013; Brazier 2015; Apraku et al. 2018a). Livestock plays an essential role in sustaining the fragile economies of most developing countries, with more than half of the rural population being sustained by livestock production (Kaasschieter et al. 1992; Assan and Kumar 2009; Assan 2014). Kaasschieter et al. (1992) note that "the largest share of the world's livestock population is found in developing countries: 61% cattle, 43% sheep, 79% goats, and 57% pigs." Of the stated livestock, "the production of small livestock like goats and pigs strengthens the capacity of communities to adapt to climate change" (Ndlovu 2010; Girum et al. 2008). Small livestock have higher chances of surviving under climate change conditions like increased droughts compared to bigger livestock since they require smaller grazing land and less amount of water and are less affected by the depletion of pastures. Assan (2014) argues that livelihoods sustainability in rural areas can be achieved through livestock production in the face of rapidly changing climate that weakens crop production especially in arid climates like Zimbabwe. It is against this backdrop that the International Union for Conservation of Nature and Natural Resources (2010) observes that rural livelihood systems in Southern Africa are converting from mixed crop-livestock to pure rangeland systems. Although livestock production in arid agroecological regions of Zimbabwe is yet to be endorsed as an agricultural policy, there is increased support for the uptake of livestock rearing from the government's Livestock Production Department (LPD).

Given the foregoing, the research thus examined the efficacy of small livestock production as a climate change adaptation strategy in Matobo District compared to the much traditionally desired cattle rearing. This switch is being pioneered by Khulasizwe which is a nongovernmental organization (NGO) operating in Matobo wards 3, 4, 10, and 19 with the aim of enhancing household livelihood security. Khulasizwe has thus far distributed goats to 380 households in the targeted wards (World Vision 2015). It is crucial to investigate community narratives (and experiences) about the effectiveness of this technical adaptation strategy. The study sought to answer the following research questions: What benefits (if any) are Matobo community members deriving from the switch to small livestock production? What has been the uptake of this strategy in the broader Matobo community over and above the project beneficiaries?

## The Impact of Climate Change in Zimbabwe

There is consensus that climate change is causing more socioeconomic, health, and environmental harm in developing countries (Van Aelst and Holvoet 2016; Kahsay and Hansen 2016; Apraku et al. 2018a, b). The harm is exacerbated by the fact that most of the population these countries is heavily reliant on the natural environment for survival, both of which are highly susceptible to climatic changes. Consequently, climate change is negatively impacting the livelihoods of the poor as well as deepening poverty in many developing countries (Africa Partnership Forum 2008; Nyong 2009; Phiri et al. 2014, 2019). In Zimbabwe, the adverse impacts of climate change are particularly felt in the rural areas where the majority of the population (67%) lives and mostly depend on agriculture-based livelihoods (ZIMSTAT 2013). Within the context of the unfolding adverse effects of climate change, vulnerable communities are adapting to these conditions by "rearing small drought resistant livestock such as goats" (Zimbabwe's National Climate Change Response Strategy 2014).

### Climate Change and Vulnerability

Climate-induced variability increases the vulnerability of rural livelihoods and reduces the ability of households to deal with risks, shocks, and stresses (Prowse and Scott 2008). Dulal et al. (2010) and Ndlovu et al. (2019) concur with the above argument that, since households typically have limited assets, they are at increased risk (exposure) and their ability to cope is restricted. Cutter et al. (2000) and Dube et al. (2017) point out that livelihood vulnerability to climate change can be usefully understood as an outcome of biophysical and social factors. Firstly, biophysical climate change vulnerability refers to the level of exposure communities' physical impacts of sea level rise, increase in sea surface, or atmospheric temperatures. Secondly, social vulnerability is explained as partially the product of those factors that shape the susceptibility of communities to harm and those that govern their ability to respond. It also includes "place inequalities" – those characteristics of communities and the built environment, such as the level of urbanization, growth rates, and economic vitality – that contribute to the social vulnerability of particular places (Cutter et al. 2000; Shah et al. 2013).

Moss et al. (2001) add a third dimension as an external assistance, which is defined as "the degree to which a region may be assisted in its attempts to adapt to change through its allies and trading partners, diasporic communities in other regions, and international arrangements to provide aid." In contrast to the United Nations (2004), this conceptualization of vulnerability includes factors outside the vulnerable system, such as characteristics of the stressor and the expected level of external assistance.

Four dimensions have been identified as defining a vulnerable situation within the literature. The first dimension consists of human-environment system, population group, economic sector, geographical region, or natural system (Downing and Patwardhan 2004). They note that other researches limit vulnerability to social systems coupled with human-environment systems (Turner et al. 2003) whereas others apply it to any system that is potentially threatened by a hazard (Mcbean 2012). The second dimension is the attribute of concern. Examples of attributes of concern include human lives and health; the existence, income, and cultural identity of a community; and biodiversity, carbon sequestration potential, and timber productivity of a forest ecosystem (Downing and Patwardhan 2004).

The third dimension is the hazard. United Nations (2004) defines a hazard broadly as "a potentially damaging physical event, phenomenon or human activity that may cause the loss of life or injury, property damage, social and economic disruption or environmental degradation." Hence, a hazard is understood as some influence that may adversely affect a valued attribute of a system. A hazard is generally but not always external to the system under consideration. For instance, a community may also be threatened by hazardous business activities or by unsustainable land management practices within this community. Hazards are often distinguished into discrete hazards, denoted as perturbations, and continuous hazards, denoted as stress or stressor (Turner et al. 2003).

Finally, the fourth dimension is the temporal reference which denotes the point in time or time period of interest. Specifying a temporal reference is particularly important when the risk to a system is expected to change significantly during the time horizon of a vulnerability assessment, such as for long-term assessments of anthropogenic climate change (Downing and Patwardhan 2004).

## Adaptation Strategies in Zimbabwe

Adaptation strategies need to be understood using the concept of "adaptive capacity." Adaptive capacity has been defined as the ability or potential of a system to respond successfully to climate variability and change, and this includes adjustments in both behavior and in resources and technologies (IPCC 2007). Other scholars define it as the potential of a system to adjust itself to change (Watts and Bohle 1993; Smit et al. 2000; Turner et al. 2003; Fussel and Klein 2006). According to Armitage and Plummer (2010: 6), adaptive capacity is:

> generally referred to as the capability of a social-ecological system to be robust to disturbance, and to adapt to actual or anticipated changes (whether exogenous or endogenous). From a social systems vantage point adaptive capacity is determined by the suite of resources (technical, financial, social, institutional, political) held, and the social processes and structures through which they are mediated (i.e. governance).

From the perspective of the social sciences, adaptive capacity does not just explain how people cope with change but reflects an ability to learn, experiment, and foster innovative solutions in complex social-ecological circumstances (Cutter et al. 2003; Armitage and Plummer 2010). Therefore, the capacity of households and communities to adapt to climate change cannot be under emphasized. In response to the negative climate change impacts, smallholder farmers are adopting a variety of adaptation strategies in Zimbabwe. There is consensus that rural communities in Zimbabwe have not been mere passive victims of climate change but have been action oriented in adapting to climate variability (Gukurume 2013; Dube et al. 2018; Kahsay and Hansen 2016). Crop diversification, mixed crop livestock systems, changing planting times, and conservation farming are some of the strategies that have been adopted by communal farmers in Zimbabwe (Munhande et al. 2013; Musarurwa and Lunga 2012; Makuvaro et al. 2014).

In addition, Mano and Nhemachena (2006) argue that climate change has severely reduced chances of rain-fed crop farming in arid regions of the country such that they suggest livestock production as the most suitable strategy that can drive communities out of food insecurity and poverty. Gukurume (2013) notes that livestock production was on the increase in Bikita and other places in dry regions, indicating that farmers are replacing crops with livestock production in order to promote livelihood sustainability. The government through its departments is also encouraging farmers in areas prone to increased dry spells to invest more on livestock, because they can withstand the harsh weather conditions induced by climate change. Small stock like goats, sheep, and indigenous poultry are becoming dominant as people try to cope with drought in areas continuously receiving little rainfall (Gukurume 2013). However, Munhande et al. (2013) observed that climate change hugely contributes to extreme temperatures, severe water scarcity, and reduced fodder for livestock, especially large stock leading to poor animal production.

Water harvesting techniques (WHTs) are another climate change adaptation strategy available to communities. Rain-water harvesting (RWH) technologies are a range of techniques used for collecting, storing, and conserving rainfall and surface runoff in arid and semiarid regions (Mutekwa and Kusangaya 2006). WHTs are more amenable to "small livestock" because they consume less water compared to "large livestock" like cattle.

A very traditional and significant adaptation strategy that had been practiced particularly for rural communities is livestock herding. According to Leichenko and O'Brien (2002) livestock herding, including nomadic pastoralism, remains one of the indigenous strategies best adapted to frequent droughts in dry land areas such as in Namibia and Botswana. Seasonal movement of livestock, splitting up of herds, changing herd composition, and distributing livestock among relatives and friends in different areas minimize risk from droughts, floods, or diseases. Appropriation of the wetter areas and water sources for cultivation and fencing inhibits these crucial strategies, however.

## Study Site and Context

The study was conducted in Matobo Rural District of Zimbabwe which lies in ecological regions IV and V characterized by low and erratic rainfall (450–650 mm/year) with very high temperatures. Of the 27 wards in Matobo District, the study wards were (Gwezha, ward 3; Marko, ward 4; Makhasa, ward 10; and Zamanyoni, ward 19). Climate change has worsened the already extreme weather patterns in the district. Drought is an example of this phenomenon that has affected Matobo District in a negative way. Data emerging from the interviewed farmers shows that the prolonged droughts have dried up natural water points such as streams, rivers, and swampy areas (Fig. 1). Consequently, their pastures and grazing lands for their livestock have been depleted. **The study was conducted from March to September 2017**.

## Materials and Methods

The research employed a qualitative methodology which was more suitable for capturing community narratives and experiences about the appropriateness and effectiveness of small livestock production within the context of a changing climate in Zimbabwe. As highlighted by Turner et al. (2003), the primary strength of the qualitative approach to cultural assessment is the ability to probe for underlying values, beliefs, and assumptions. Data collection tools employed included the use of in-depth semi-structured interviews (50 households), 5 focus group discussions, and 5 key informants conducted using purposive sampling in 4 selected wards of Matobo District of Zimbabwe. The limitations of the study are that findings on how small-holder farmers adapt to climate change through small livestock in Matobo District cannot be generalized to other districts in Zimbabwe and beyond. Success of these

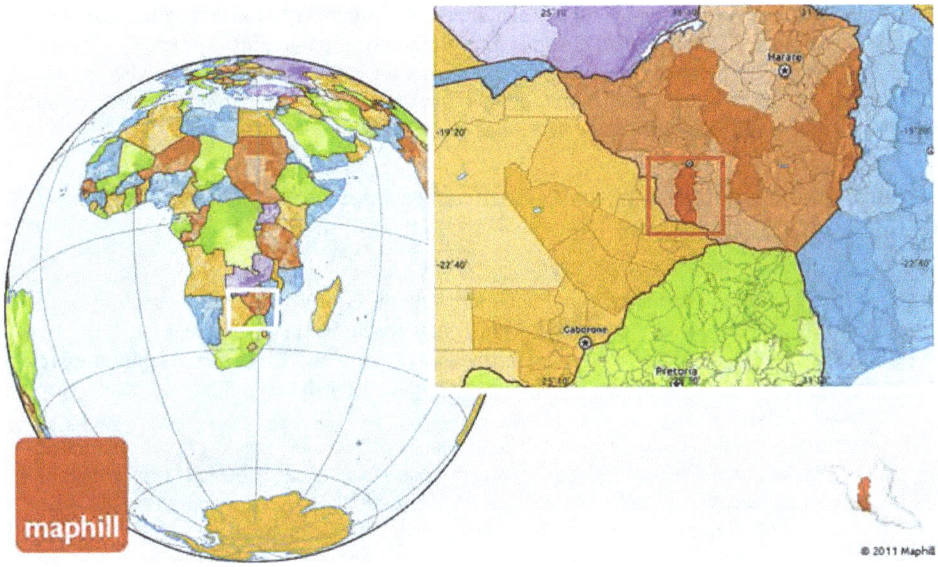


**Fig. 1** Map of Zimbabwe (showing Matobo District)

technical climate change adaptation strategies is location specific. However, important lessons can be drawn from the narratives of the smallholder farmers in semiarid districts and go on to influence policy.

## Results and Discussions

### Small Livestock and Climate Change Nexus in Matobo

Small livestock, namely, sheep and goats, are relatively more adaptive to negative changing climatic conditions compared to cattle. Goats are both grazers and browsers, while cattle are limited to the latter (Girum et al. 2008). The chances of survival for small livestock are higher compared to cattle in the context of a changing climate. Moreover, on the basis of their size, the goats consume less from the pastures in a specified area in comparison to cattle that demand more feed due to their sheer size. The amount of water drunk by goats and sheep is far less than that of cattle. Small livestock have a comparative risk aversion advantage in that respect. The suitability and appropriateness of small livestock in a climate change environment are premised on the aforementioned characteristics.

#### Goat Production

In the context of the study area, a key informant justified the rearing of small livestock when he explained that:

> Even in times of drought, small stock (goats) herd will multiply because they reproduce at two to three times the rate of cattle. A goat may produce one to four kids in a year as opposed to one birth every one and a half years to two years for cattle. The farmers in Matobo can attest to that. It is a known fact in Matobo that goatherders fare well in comparison to cattle herders in general. (Key Informant 4, Khulasizwe Trust)

Furthermore, another important statement was made in respect to small livestock rearing in Matobo. Key informants 1 and 3 indicated that:

> Small livestock are particularly important and useful during and after a drought or disaster period. It is easy for farmers who are keeping goats for instance, to sell the goats for cash, give goats as a gift, or slaughter for food or traditional ceremony. This is because the amount of money and food generated by a single goat is optimal for day to day transactions. (Key Informant 3, AGRITEX)
>
> The most popular small livestock that the farmers keep are goats. Pigs are shunned because of religious reasons…and generally because they need to be fed a lot which is highly cost ineffective and therefore not sustainable particularly for rural communities who are already struggling enough to meet household needs. The sheep herd is relatively small as well because there is too much work involved in rearing them. Sheep need special care. (Key informant 1, World Vision)

Similarly, the views raised by interviewed smallholder farmers were in tandem with what was also raised by the key informants.

Three of the respondents from different wards commented that:

> Small livestock are a good insurance against crop failure. When our harvest is low– we have an easy alternative that is of slaughtering a goat. We value our goats because they give us meat, milk and manure for our gardens. We also use the goat skin as a carpet and sell to obtain cash for petty trading. (Household Respondent 26, Marko)
>
> Goats are doing very well for me despite the heavy rainfalls and hot temperatures this year. I started with 5 goats but I now have 27 goats in less than 3 years. They multiply faster compared to cattle provided that they are dosed regularly. I love rearing goats because they do not demand stock feed like cattle. Moreover, they can even feed on groundnuts and round nuts leaves. Goats are basically easy to maintain in a climate change environment. (Household Respondent 33, Makhasa)
>
> I do not have a lot of goats myself but I have friends who are rearing a lot of them. I know of one farmer who possesses over 50 goats. From my observation, these small livestock don't collapse and die due to high temperatures and drought. They survive unlike cattle. They feed close to the homestead and can easily be watered by the owners. In times of crisis, goats can be exchanged for money or grains for food or feeding the community during a funeral wake. (Household Respondent 21, Zamanyoni)

A key informant from AGRITEX emphasized that even those who do not have cattle perceive goats and sheep as a symbol of wealth. One of the commentators in a focus group discussion indicated that there is a lot of contempt in society for one that does not have any kind of livestock and few villagers are likely to attend his funeral. For most farmers who do not have cattle, goats are a good source of relish during funerals when villagers gather in the bereaved homestead. The key informant added that small livestock are also kept for security reasons. They are likely to cushion the farmers during times of drought when the harvest from their fields is very low. They can be eaten or exchanged for grain or sold for cash to buy other household consumables during hard times. This point was also raised by interviewed small-holder farmers earlier in the discussion. It is clear from the above comments that small livestock, specifically goats, play a pivotal role as a livelihood strategy in Matobo District.

The smallholder farmers are aware of the competitive advantage of rearing goats over cattle in their specific area of habitation. Firstly, as a source of food, goats are easily slaughtered to obtain meat to meet food needs at the household level or feeding community members in funeral wakes and traditional and cultural rituals like weddings. Goat milk is also a preferred source of food in selected households specifically for children. Secondly, goats are acknowledged to be easy to keep and maintain in a harsh climatic environment compared to cattle because they consume less food and are more drought resistant compared to cattle. Thirdly, goats are utilized by farmers as a medium of exchange to pay school fees, hospital bills, and other urgent needs of the households. Interviewed farmers unanimously indicated that it is easier to sell goats than cattle.

It emerged from the sampled respondents that the Matabele goat is common in Matobo District because it is an indigenous breed that has been kept there since time immemorial. However, the key informant from AGRITEX indicated other breeds like the Boer goat and Kalahari red are common in areas like Gwanda and Beitbridge. This is so on account of farmers' cross-breeding them due to the proximity and accessibility of South Africa at the Limpopo province. The Mashona goat is mainly confined to Mashonaland province of Zimbabwe (Table 1).

Experts in the field of small livestock rearing concur that goats are an effective strategy for adapting to climate change in the study area. Their views go beyond the perception of interviewed farmers who are yet to fully appreciate this knowledge beyond merely keeping goats as a tradition. The key informant from Khulasizwe justified the rearing of small livestock like goats over cattle for a number of reasons. He noted that:

> Goats can withstand the harsh climatic changes and are not prone to collapsing like cattle. They feed on vegetation not used by other domestic species of animals since they are browsers. Goats do not easily succumb to diseases. Farmers living in marginal agricultural areas can keep them. Moreover, they are easy to handle and can be looked after by children and women. (Key Informant 4, Khulasizwe)

Another key informant from the Department of Livestock Production and Development gave a lengthy response in showing the viability and effectiveness of goats' production over cattle production. He narrated that:

> Goats multiply very easily and flocks expand in a relatively short time forming a major part of the family asset base in a harsh changing climate. Goats have an economic advantage over large ruminants because they are cheap to buy hence they provide an opportunity for vulnerable people to own them. They can quickly and easily be sold off for cash to meet family emergency needs like health costs, educational needs, in times of death and therefore are an easily disposable asset during times of poverty and calamities. With goat rearing, there are none or few social, religious and cultural taboos associated with keeping them in Zimbabwe. (Key Informant 5, DLPD)

**Table 1** Common goat breeds in Zimbabwe

| Goat breed | Characteristics |
|---|---|
| Matabele goat | This breed is found mostly in the southern part of the country. It is large framed animal |
| | Mature weight can reach up to 50 kg plus for males and between 35 and 40 kg females |
| | It is a dual-purpose breed (meat and milk production) |
| | Has multiple colors ranging from white through brown, mixed to black |
| | Can give birth to singles, twins, and triplets |
| | Adapted to local conditions |
| | Gestation period is between 145 and 154 days |
| | Can be mated for the first time at 18 months of age |
| Mashona goat | It is small-framed breed |
| | It is highly prolific |
| | Tolerant to local diseases |
| | Found in the northern parts of the country |
| | Mature weight is 30 and 35 kg for females and males, respectively |
| | Color ranges from white, mixed to black |
| | Edible and saleable proportion after dressing is 48% of the live body weight (equivalent to 12–14.5 kg) |
| | Sexual maturity is 7 months but a relatively slow growth rate. Goats are mainly kept for meat |
| | They have good potential for selection to increase productivity |
| Boer goat | It originates from South Africa and is well adapted to the southern parts of the country |
| | This is a large framed breed, with white body and a brown neck and head |
| | Adult female can grow up to 60–70 kg and males up to 70–90 kg |
| | It is mainly for meat production |
| | Very stocky body, well-muscled, strong bones |
| | Ears are pendulous and horns are prominent |
| | Very prolific breed, kidding rate of 1, 5 (three times in 2 years) |
| | 50% twins and 7% triplets |
| | Valuable for meat, milk, and skin but has been promoted for its meat potential |
| Saanen goat | Originated from Switzerland or France |
| | Large size, white color with black spots on the nose, ears, and udder |
| | Hair coat is short; ears are erect and forward and upward |
| | Face is straight, slightly dished |
| | Good body conformation |

Source: Khulasizwe File Report (2018)

From a technical point of view, it was evident that small livestock especially goats are a viable adaptation strategy to climate change effects. Key issues raised in support of their breeding were by virtue of their size in comparison to cattle; the risk of loss is small. It was suggested that for every cow one has an equivalent of six goats. The land size for communal farmers is small, and therefore keeping goats rather than cattle enables them to reach economies of scale. The eating habits of goats (grazers and browsers) mean they have a more varied diet and can survive in drier areas. They basically can survive in any terrain that can either be hilly or flat terrain. In addition, it was revealed by the key informants that where water is scarce, goats can survive without drinking water for 2 days. However, the conscious decision to rear small livestock for the purpose of responding to climate change is yet to be understood by the smallholder farmers depending on efforts by government and its partners to spread that knowledge and the willingness of the respondents to accept that reality.

## Sheep Production

Data gathered from the study area also revealed that 20% of the interviewed farmers rear sheep. However, despite their resilience to the impact of climate change, they are not produced en masse in Matobo District in comparison to goats and chickens. The reasons given by two interviewed households were that:

> Sheep easily get lost because they can hardly re-track their way back if they feed a distance away from home. Due to that fact, sheep demand more labour than other livestock since they require close attendance almost all the time. (Household Respondent 24, Gwezha)
>
> Keeping sheep is hard work and not ideal for these reasons. When thieves and predatory animals like jackals and hyenas pounce in the night, they do not make any noise. We are not alerted that there is impending danger the same way goats make a loud noise when disturbed. As a result, we have lost quite a number of sheep due to this problem. We are therefore, reluctant to rear them in our village. (Household Respondent 40, Zamanyoni)

On a different note, the interviewed farmers who are rearing sheep had other reasons that they gave for doing so besides their adaptability to climate change. The reasons are purely religious beliefs. The following views confirm that:

> We keep sheep for the purpose of enterprising especially with white garment churches known as apostolic sects. They believe that the tail of the sheep on which they extract fat brings good luck. Whether that is true or not is not an issue for me. As long as I can make money out of them, I do not care. I am keeping 53 sheep. It's a big market for me and my household. (Household Respondent 3, Marko)
>
> Most of those households that keep sheep are superstitious. They believe that if you rear sheep your homestead can never be struck by lightning. Climate change has also come about with lightning and terrifying thunder. As a protection measure, sheep are kept to ward off that risk. For sure, I know no household in the village that has been struck by lightning that kept sheep. (Household Respondent 19, Makhasa)

The key informants acknowledged that specific farmers in the district held such indigenous knowledge beliefs about sheep being said to prevent lightning and the supernatural powers they possess to ward off bad luck and bring good fortune. Their belief system primarily informs their choice of rearing sheep. The production of sheep for the purpose of adapting to climate change is secondary under those circumstances. In this instance, small livestock keeping is a tradition passed on through generations and is not intended as a direct response to climate change. All household respondents kept one form of small livestock or another. However, it should be noted that it is not attributed to responding to climate change per se but a continuation of what has always been the norm in the district since time immemorial. The findings also show that because of the traditional inclination to goat rearing, farmers' knowledge on which types of goats are reared is lacking and immaterial. Farmers in Matobo are not really concerned about the types of breeds that they are rearing; they keep whatever breed that they can access or obtain. This lack of knowledge on breed types has however an implication on the effectiveness of small livestock as adaptation options to climate change. Sustained and successful adaptation strategies demand a level of knowledge and understanding about livestock breeds and their resilience to the adverse effects of climate change.

### Feminization of Small Livestock Rearing in Matobo

In Matobo, rearing of sheep and goats is mostly relegated to women. While men do keep small livestock in Matobo, it is understood that it is a feminine exercise. On that basis, men play a peripheral role in small livestock rearing and would prefer to focus on owning cattle. It is an unwritten law that women (especially those that are married) may not slaughter a beast under any circumstances without prior approval of their husbands. This points to a strong patriarchal culture that is predominant even in climate change adaptation strategies.

## Constraints to Small Livestock Rearing

Despite the fact that small livestock have a comparative advantage in adapting to climate change over cattle in the study area, the production is beset by subtle challenges that threaten its viability and effectiveness. For instance, the key informant from the DLPD commented that:

> While goats' production is a good option for adapting to climate change in Matobo, they have their own share of problems. Goats can be destructive animals only if overgrazing in areas occurs, resulting in reduced or loss of tree regeneration. Moreover, goats have a tendency to stray and eat neighbours' crops. Therefore, goats should be herded properly. (Key Informant 5, DLPD)

The respondents in the study area were more concerned about the security of their livestock more than anything else. Thieves and predatory animals like hyenas and jackals threatened the viability of their small livestock. The following comments were made during a focus group discussion:

> Jackals are a big problem in our village. Four jackals can chase one goat till they catch and kill it. Sometimes we see the jackals during the day. They are so daring these days. I recently lost 3 of my goats as a result of jackals. Unless, one devises a mechanism of herding them throughout the day– they will get eaten. I am actually better off– my neighbor lost 12 goats within a month. Jackals are really a menace. (Marko Focus Group Discussion, Participant 2)
>
> Stock theft is a common occurrence in our village. The incidences are rampant probably because of hunger perpetuated by what you call climate change. Crop yields are no longer enough to feed the communities throughout the year. Sadly, some people in the village resort

> to stealing chickens and goats to make ends meet. I own 65 goats. I lost 13 just last year to thieves. It's not acceptable. (Zamanyoni Focus Group Discussion, Participant 9)
>
> Hyenas have cost me a great deal of my goats. They target the kid goats especially. Twenty-one of my goats got eaten by hyenas. We cannot even hunt them down because it is a criminal offence. The government is not doing anything to protect our goats from the menacing hyenas. (Gwezha Focus Group Discussion, Participant 5)

The interviewed farmers also narrated that thieves are so daring in certain instances. It was noted that thieves can even drive goats in broad daylight when they have observed that they are not herded by any one. The villagers suspect that they sell them in nearby urban areas like Bulawayo where there is a ready meat market. The key informants from World Vision, Khulasizwe, AGRITEX, and DLPD confirmed the issues raised by farmers on losses of their small livestock as a result of thieves and predatory animals. As a result, some livestock security strategies have been developed by community members. Plate 1a, b is a new idea of goat and sheep kraals in Matobo District.

**Plate 1** (**a** and **b**) Goat and sheep kraals

The key informant from DLPD commented that they are encouraging the construction of these kraals because:

> . . .they are constructed in an elevated manner to prevent hyenas and jackals from killing the sheep and goats. They also serve the purpose of protecting the sheep from harsh weather conditions like intense heat and heavy rainfall. Traditional kraals which are constructed at ground level were not as secure as this new model. Thus, farmers are able to protect their livestock from being attacked and also prevent diseases caused by exposure from the heat and intense precipitation. (Key Informant 5, DLPD)

The Department of Livestock Production and Development in conjunction with local NGOs has been instrumental in encouraging farmers in the study area to construct these new kraals to preserve and protect their small livestock.

Another very important challenge that negatively affects the interviewed farmers is the diseases that reduce the number of their small livestock. Diseases like liver fluke, round worms, aching eyes, lump skin, pulp kidney, and heart water have affected livestock in the study area in varying degrees. In the face of these challenges, interviewed farmers seem to be helpless. There is little that they have done to treat their animals despite the few trainings received from the veterinary department. Farmers do not seem to understand the kind of diseases that are affecting their livestock. When asked about what kind of diseases affect their livestock, the farmers usually respond by describing what happened; they fail to state the disease that has led to the mortality of their livestock either in vernacular or in English. They just refer to them as *igazi* (*blood*), *or* any other terms that describe the symptoms not the disease. Three interviewed farmers commented that:

> Out of 35 goats, I lost all 16 in a single year through a strange disease, but could do nothing about it. I just watched helplessly. Some of us don't believe that drinking stagnant water can affect our livestock. Our goats used to drink from the same rivers and dams which are said to be causing their death. We don't understand why. We don't have money to buy the chemicals to treat our animals. (Household Respondent 41, Makhasa)
>
> We have received training on how to identify diseases and cure them from the Veterinary Department. The medication for dosing and injecting our goats are both expensive and not accessible in Maphisa. Bulawayo is too far where we are referred to. We end up resorting to traditional medicines to cure our livestock. The Veterinary Department discourages us from doing so. But what other options do we have? (Household Respondent 27, Zamanyoni)
>
> Goats and sheep are affected by liver fluke and bowels due to drinking stagnant water. There are no signs or symptoms for these two diseases, one just discovers after the death of the animal. I have lost three goats because of liver fluke. (Household Respondent 15, Gwezha)

Over and above the lack of disease knowledge, the expensive medication from the veterinary department is also responsible for the high mortality rate of goats and sheep. The farmers are too poor to purchase the recommended medication to cure their animals. Consequently, they suffer heavy losses due to diseases like liver fluke, round worms, aching eyes, lump skin, pulp kidney, and heart water. Farmers were clueless on how some of those diseases could be dealt with. Moreover, there is some resistance on dipping goats which solves the tick problem. Farmers are used to dipping cattle. They are reluctant to dip goats and sheep. That attitude allows their livestock to easily succumb to tick borne diseases.

The views of the respondents concur with other studies which also show that communities in Africa grapple with what are called multiple stressors (Adger et al. 2003; Osbahr et al. 2008; Drimie and Gillespie 2010). Multiple stressors (in this case predatory animals and thieves) are other challenges that come about or are in existence when communities are attempting to adapt to climate change. These multiple stressors complicate the ability of farmers to adapt to climate change smoothly. The AGRITEX official summarized the challenges faced by interviewed farmers when he remarked that:

> There are a number of challenges faced by the small livestock farmers ranging from diseases, thieves, predators and lack of finance to buy treatment for their livestock. For most of the constraints that they face, nothing has been done much to help them. For fear of retribution through current legislation, they watch as their livestock are mauled by the hyenas and jackals. They have not taken meaningful steps to address the diseases that affect their livestock. Thieves sometimes help themselves to their livestock and for those that are able to keep significant numbers they find themselves with a dilemma of markets to sell their livestock. (Key Informant 1, AGRITEX)

He however further suggested that the attitude of farmers also needs to be worked on. They do not classify themselves as goat farmers. They do not keep records. Livestock production is something to invest time and resources on. Few are taking steps to exploit the huge potential that lies with small livestock production. It was also apparent that interviewed farmers rear goats without an awareness of the type of breed which they keep. For all the interviewed farmers, a goat is appreciated just because it is a goat, what breed it is does not matter.

## Institutional Mandate Response on Small Livestock Production

The Livestock Veterinary Department in partnership with AGRITEX has been very instrumental in the study area in promoting small livestock rearing. The Livestock Veterinary Department offers technical support to the smallholder farmers at no charge to assist them to improve on small livestock production. Two household respondents from different wards remarked that:

> The veterinary department has been very helpful to us as farmers in our village. They conduct training and workshops to teach us on small livestock breeding. We have been taught that are goats must be dosed and well-sheltered so that they are not exposed to too much heat and precipitation. Our goats are now dipping although some farmers still find it strange to do that because we grew up knowing it is cattle that dip. They respond fast even when there is an outbreak of a disease that affects our chickens as well. (Household Respondent 34, Makhasa)
>
> I am one of the few farmers that have benefited from teachings facilitated by the Veterinary Department. I was trained at Maphisa to be a paravet for a week. Because of my passion in goat rearing I have over 50 goats now from the mere 14 I had before I was

> trained. All my goats go for dipping every month. I dose them regularly. I have thus constructed this new kraal to prevent them from being mauled by hyenas and jackals. (Household Respondent 50, Zamanyoni)

However, another research participant in the study area had a different view on the role of the veterinary department. He exclaimed that:

> Our Veterinary personnel are not well capacitated in our district in terms of resources to serve the farmers. They have lots of literature in their office but they have no transport to go around distributing the literature and equipping farmers as they should. They do not even have up to date statistics because they don't have the means to get to where the people are. (Household Respondent 22, Gwezha)

It was clear from the 67% interviewed farmers that the Livestock Veterinary Department is making great strides in capacitating smallholder farmers to enhance the production of small livestock in the study area. They acknowledged that they have received information through field visits on issues relating to small livestock dipping, dosing, climate friendly shelter, and dealing with diseases. However, 33% of the interviewed farmers felt that the department is limited by resource constraints from the central government. They argued that the technical experts from the veterinary rarely do field visits and that they only avail themselves during emergency situations such as disease outbreaks threatening small livestock. They also claimed that they occasionally come when they have a joint program with NGOs like World Vision. Given such a scenario, smallholder farmers are forced to be initiative and deal with some diseases using traditional medicine or fork out money to travel from their village to Maphisa Centre to get technical advice. There was also a concern that even if one travels to Maphisa, medicine is not readily available, and therefore farmers are compelled to further go to Bulawayo which is difficult due to financial constraints.

Khulasizwe (local NGO) which used to operate in the study area from 2009 to 2013 played a crucial role in the promotion of goat production. They mainly distributed three female goats per household to promote food security and create a market for Bulawayo butcheries where the demand for goat meat was high. The researcher identified two beneficiaries who had benefited from the Khulasizwe project. They commented that:

> Khulasizwe gave me 3 goats in 2009. They also gave me and other beneficiaries cement, asbestos to build shelter for goats. Khulasizwe offered a 50/50 deal where you also paid half prize of stock and they paid the other half. They offered vaccines. This has improved livestock production in Matobo. I currently own 27 goats from the original 3 I received from Khulasizwe. (Household Respondent 49, Makhasa)
>
> Goats have a huge potential of saving not only Matobo farmers but the whole nation from a potential disaster caused by climate change. There is therefore need for the government to help the small livestock farmers to go commercially and to produce for the nation. A little more of the government presence on the ground can make a huge difference. (Household Respondent 50, Makhasa)

## Conclusions

Farmers in the study area are agreed that goats and sheep have increased their adaptive potential in the face of adverse climatic condition. Unlike cattle, the survival rates of small livestock is high in the drought prone areas like Matobo District and have proved more viable and effective. Most households from the sample kept one form of small livestock irrespective of the knowledge on the breeds that they kept. The success and viability of small livestock production largely depend on the input of government and other stakeholders like the Department of Livestock Production and Development, Jairos Jiri, World Vision, and Khulasizwe working closely with farmers. Their contribution to farmers through expert training and provision of small livestock is key to successful adaptation. Therefore, that working relationship should be strengthened and maintained.

## Recommendations

Based on the above findings, this study recommends the following:

The government, at local and national level, can facilitate and implement adaptation strategies through policies that support farmers and make resources available. The government has documented action plans and strategies to tackle the challenges of climate change through the Ministry of Environment, Water and Climate. However, there is need for government to "walk the talk" and implement what is documented on paper and implement it on the ground. There is a discrepancy over what the government intends to do and what communities experience in their environment.

There is need for government and NGOs like AGRITEX, EMA, Department of Livestock Production and Development, World Vision, Khulasizwe, and Jairos Jiri to coordinate their activities. Research participants noted overlaps in how such programs are implemented. Coordination among them assists in avoiding or at least reduces overlap of activities. Climate change adaptation strategies if coordinated well can help in discovering of existing gaps in need of implementation.

Adoption of technical adaptation strategies like small livestock rearing should be broadened through knowledge dissemination and awareness in all wards. The challenge currently is that interventions target a few farmers and leave out others. Going forward, knowledge and training should be emphasized until communities fully appreciate the adaptation options suitable and appropriate for their areas. Raising awareness of climate change issues can be done through the local media and extension staff. Dissemination of climate change information from national and local meteorological stations to farmers through extension agents is also very critical.

## References

Adger WN, Huq S, Brown K, Conway D, Hulme M (2003) Adaptation to climate change in the developing world. Prog Dev Stud 3(3):179–195

Africa Partnership Forum (2008) Climate challenges to Africa: a call for action. Available from: http://www.africapartnershipforum.org

Apraku A, Moyo P, Akpan W (2018a) Coping with climate change in Africa: an analysis of local interpretations in Eastern Cape, South Africa. Dev South Afr. https://doi.org/10.1080/0376835X2018.1482199

Apraku A, Akpan W, Moyo P (2018b) Indigenous knowledge, global ignorance? Insights from an Eastern Cape Climate Change Study. S Afr Rev Sociol 49(2):1–21. https://doi.org/10.1080/21528586.2018.1532813

Armitage D, Plummer R (2010) Integrating perspectives on adaptive capacity and environmental governance. In: Adaptive capacity and environmental governance. Springer, New York, pp 1–19

Assan N (2014) Goat production as a mitigation strategy to climate change vulnerability in semi-arid tropics. Sci J Anim Sci 3(11):258–267

Assan JK, Kumar P (2009) Introduction: livelihood options for the poor in the changing environment. J Int Dev 21(3):393–402. https://doi.org/10.1016/j.foodpol.2008.01.003

Brazier A (2015) Climate change in Zimbabwe facts for planners and decision makers. Konrad Adenauer Stiftung, Harare

Chaudhury M (2017) Strategies for reducing vulnerability and building resilience to environmental and natural disasters in developing countries. World Resources Institute, United Nations Department of Economic and Social Affairs, Division of Social Policy and Development, New York

Cutter SL, Mitchell JT, Scott MS (2000) Revealing the vulnerability of people and places: a case study of Georgetown Country, South Carolina. Ann Assoc Am Geogr 90(4):713–737

Cutter SL, Boruff BJ, Shirley WL (2003) Social vulnerability to environmental hazards. Soc Sci Q 84(2):242–261. https://doi.org/10.1111/1540-6237.8402002

Downing TE, Patwardhan A (2004) Assessing vulnerability for climate adaptation. In: Lim B, Spanger-Siegfried E (eds) Adaptation policy frameworks for climate change: developing strategies, policies, and measures. Cambridge University Press, Cambridge

Drimie S, Gillespie S (2010) Adaptation to climate change in Southern Africa: factoring in AIDS. Environ Sci Pol 13:778–784

Dube T, Phiri K (2013) Rural livelihoods under stress: the impact of climate change on livelihoods in South Western Zimbabwe. Am Int J Contemp Res 3(5):11–25

Dube T, Moyo P, Intauno S, Phiri K (2017) The gender-differentiated impacts of climate change on rural livelihoods labour requirements in Southern Zimbabwe. J Hum Ecol 58(1–2):48–56

Dube T, Mlilo C, Moyo P, Ncube C, Phiri K (2018) Will adaptation carry the future? Questioning the long-term capacity of smallholder farmers' adaptation strategies against climate change in Gwanda District, Zimbabwe. J Hum Ecol 61(1–3):20–30

Dulal HB, Brodnig G, Thakur HK, Green-Onoriose C (2010) Do the poor have what they need to adapt to climate change? A case study of Nepal. Int J Justice Sustain 15(7):621–635

Fussel HM, Klein RJT (2006) Climate change vulnerability assessments: an evolution of conceptual thinking. Clim Chang 27:301–329. https://doi.org/10.1007/s10584-006-0329-3

Girum G, Degefa T, Getachew G (2008) Risk perception and coping strategies among the Karrayu pastoralists of upper Awash Valley, Central Ethiopia. Nomadic People 12:93–107

Gukurume S (2013) Climate change, variability and sustainable agriculture in Zimbabwe's rural communities. Russ J Agric Socio-Econ Sci 2(14):89–93

Intergovernmental Panel of Climate Change (2007) Climate change 2007: impacts, adaptation and vulnerability. Working group II contribution to the Fourth Assessment Report of the Intergovernmental Panel on Climate Change. Cambridge University Press, Cambridge

International Union for Conservation of Nature (2010) Building climate change resilience for Africa livestock in sub-Saharan Africa. World Initiative for Sustainable Pastoralism, Nairobi

Kaasschieter GA, de Jong R, Schiere JB, Zwart D (1992) Towards sustainable livestock production in developing countries and the importance of animal health strategy therein. Vet Q 14(2):66–75
Kahsay AG, Hansen LG (2016) The effects of climate and adaptation policy on agricultural production in Eastern Africa. Ecol Econ 121:54–64
Khulasizwe Small Livestock Production in Matobo District (2018) Quartely Report. Unpublished
Leichenko RM, O'Brien KL (2002) The dynamics of rural vulnerability to global change: the case of Southern Africa. Mitig Adapt Strateg Glob Chang 7:1–18
Makuvaro V, Walker S, Munodawafa A, Masere TP, Murewi C, Chagonda I (2014) An overview of current agronomic practices of smallholder farmers in semi-arid Central and Western Zimbabwe. Afr J Agric Res 9(35):2710–2720
Mano R, Nhemachena C (2006) Assessment of the economic impacts of climate change on agriculture in Zimbabwe: a Ricardian approach. CEEPA discussion paper no. 11. Centre for Environmental Economics and Policy in Africa, Pretoria
McBean GA (2012) Integrating disaster risk reduction towards sustainable development. Environ Sustain 4:1–6
Moss RH, Brenkert AL, Malone EL (2001) Vulnerability to climate change: a quantitative approach. Technical report PNNL-SA-33642, Pacific Northwest National Laboratories, Richland
Munhande C, Mapfungautsi R, Mutanga P (2013) Climate risk management: actors, strategies and constraints for smallholder farmers in Zimbabwe: a case study of Chivi District. J Sustain Dev Afr 15(8):240
Musarurwa L, Lunga W (2012) Climate change mitigation and adaptation: threats and challenges to livelihoods in Zimbabwe. Asian J Soc Sci Humanit 1(2):25–32
Mutekwa V, Kusangaya S (2006) The contribution of rainwater harvesting techniques to rural livelihoods in Zimbabwe: the case of Ngundu Ward in Chivi District. Bull Anim Health Prod Afr 32(3):437–452
Nandhini DU, Suganthi S (2018) Livestock importance in organic farming. Approaches in Poverty, Dairy and Veterinary Sciences 5(1). https://doi.org/10.31031/APDV.2018.05.000602
Ndlovu L (2010) Food, nutrition and health. In: Swanepoel FJC, Stroebel A, Moyo S (eds) The role of livestock in developing communities: enhancing multifunctionality. CTA, Wageningen
Ndlovu S, Mpofu M, Moyo P (2019) Debunking the effectiveness of in-kind transfers in alleviating urban household food insecurity in Bulawayo, Zimbabwe. Dev South Afr. https://doi.org/10.1080/0376835X.2019.1584031
Nyong A (2009) Climate change impacts in the developing world: implications for sustainable development. In: Brainard L, Jones A, Purvis N (eds) Climate change and global poverty: a billion lives in the balance? Brookings Institution Press, Washington, DC
Osbahr H, Twyman C, Adger WN, Thomas DSG (2008) Effective livelihood adaptation to climate change disturbance: scale dimensions of practice in Mozambique. Geoforum 39(6):1951–1964
Phiri K, Ndlovu S, Chiname TB (2014) Climate change impacts on rural based women: emerging evidence on coping and adaptation strategies in Tsholotsho, Zimbabwe. Mediterr J Soc Sci 5(23):2545–2552
Phiri K, Dube T, Moyo P, Ncube C, Ndlovu S (2019) Small grains "resistance"? Making sense of Zimbabwean smallholder farmers' cropping choices and patterns within a climate change context. Cogent Social Sciences 5(1):1622485
Prowse M, Scott L (2008) Assets and adaptation: an emerging debate. IDS Bull 39(4):42–52
Shah KU, Dulal HB, Johnson C, Baptiste A (2013) Understanding livelihood vulnerability to climate change: applying the livelihood vulnerability index in Trinidad and Tobago. Geoforum 47:125–137
Smit B, Burton I, Klein RJT, Wandel J (2000) An anatomy of adaptation to climate change and variability. Clim Chang 45:223–451
Turner BL, Kasperon RE, Matson PA, McCarthy JJ, Corell RW, Christensen L, Eckley N, Kasperson JX, Luers A, Martello ML, Polsky C, Pusipher A, Yauch CA, Steudel HJ (2003) Complementary use of qualitative and quantitative cultural assessment methods. Organ Res Methods 6(4):465–481

United Nations (2004) Living with risk: a global review of disaster reduction initiatives. United Nations International Strategy for Disaster Reduction, Geneva

Van Aelst K, Holvoet N (2016) Intersections of gender and marital status in accessing climate change adaptation: evidence from rural Tanzania. World Dev 79C:40–50

Watts MJ, Bohle HG (1993) The space of vulnerability: the causal structure of hunger and famine. Prog Hum Geogr 17(1):43–67

World Vision (2015) Goat Rearing in Matobo District, Zimbabwe, Annual Report. Unpublished

Zimbabwe's National Climate Change Response Strategy (2014) Government of Zimbabwe Ministry of Environment, Water and Climate: annual report. Institute of Environmental Studies, Harare

ZIMSTAT (2013) Poverty and poverty datum line analysis in Zimbabwe 2011/12. ZIMSTAT, Harare

# 2

# Biochar for Climate Change Adaptation: Effect on Heavy Metal Composition of *Telfairia occidentalis* Leaves

Doris Akachukwu, Michael Adedapo Gbadegesin, Philippa Chinyere Ojimelukwe and Christopher John Atkinson

## Contents

Introduction ....
Materials and Methods
- Sociodemographic Data
- Biochar Preparation
- Soil Preparation and Biochar Application
- Procurement, Cultivation, and Growth Indices of *Telfairia occidentalis*
- Heavy Metal Determination
- Data Analysis
- Results and Discussion

Conclusion
References

---

D. Akachukwu (✉)
Department of Agriculture, Health and Environment, Natural Resources Institute, University of Greenwich, Chatham, UK

M. A. Gbadegesin
Department of Biochemistry, Faculty of Basic Medical Sciences, University of Ibadan, Ibadan, Oyo State, Nigeria

P. C. Ojimelukwe
Department of Food Science and Technology, Michael Okpara University of Agriculture, Umudike, Abia State, Nigeria

C. J. Atkinson
Natural Resources Institute, University of Greenwich, London, UK

Department of Agriculture, Health and Environment, Natural Resources Institute, University of Greenwich, Chatham, UK

**Abstract**

Gas flaring is a key contributor of greenhouse gases that causes global warming and climate change. Adaptation measures for tackling impacts of climate change have gained much research interest. This chapter assessed vegetable farmers' perception of gas flaring and the effect of biochar remediation on the heavy metal composition of cultivated *Telfairia occidentalis*. A gas-flared area, Ohaji/Egbema L.G.A of Imo State, and a non-gas-flared area, Umudike, Ikwuano L.G.A, were selected for this research. Structured questionnaire was used to elicit information from 120 respondents. Soils were collected from the study sites and transported to the greenhouse. Five different rates, 0 t $ha^{-1}$, 7.1 t $ha^{-1}$, 13.9 t $ha^{-1}$, 20.9 t $ha^{-1}$, and 28.0 t $ha^{-1}$, of palm bunch biochar were applied to the soils in plastic buckets. After 2 weeks of mineralization, two viable seeds of *Telfairia occidentalis* were planted in each bucket and watered every other day for 8 weeks. The result revealed that 63% of vegetable farmers where female, while 37% were male in the gas-flared area. A total of 97% of the farmers had knowledge of gas flaring. A total decrease of 55% percent income, 90% yield, and 67% market quality of vegetable farmers was attributed to gas-flared activities. The plant height of cultivated vegetables increased every 2 weeks with greater increase in the test plant. Heavy metal concentration (Pb, and Cr) decreased with increasing biochar rate and was significantly lower for 28.0 t $ha^{-1}$. Biochar can enhance soil fertility and help immobilize heavy metals. The effect of biochar application on the heavy metal composition is dependent on the rate of application. Biochar use could be a cheap adaptation measure in the face of a changing climate.

**Keywords**

Biochar · Climate change adaptation · Gas flaring · Heavy metal · *Telfairia occidentalis*

## Introduction

Natural gas is a component of petroleum that is released during crude oil exploitation (Mohammed et al. 2016). This natural gas is usually vented or flared (burnt off) (Mohammed et al. 2016). Venting has to do with a direct release of natural gas into the atmosphere without burning, while gas flaring is the indiscriminate burning of natural gas during crude oil exploitation. This practice is common in most countries of the world especially in developing countries like Nigeria where oil companies fail to acquire the needed infrastructure for trapping, storage, and use of natural gas. Gas venting and flaring are responsible for enormous waste of nonrenewable resource, pollution, and an annual release of greenhouse gases which contributes immensely to global warming and climate change (Ajugwo 2013). In 2017, World Bank Global Gas Flaring Reduction Partnership reported that more than 140 billion cubic meters (bcm) of natural gas was flared. This huge volume of gas can serve as a reliable source of energy for the African continent yearly, but on the contrary, it has remained an unprecedented source of pollution (Emam 2016; Ismail and Umukoro 2012). As commercial deposits of crude oil was discovered in Nigeria, the nation drifted from agriculture to oil exploitation as its national income base. Oil industry became the nation's gold mine as it accounts for over 80% of her earnings (Ohimain 2013a, b). The oil companies in the Niger Delta region of Nigeria provided more than 80% of funds used for allocating national budget (Ede and Edokpa 2015). Nigeria is ranked one of the ten countries of the world with abundant natural gas reserve (Orlando 2006). It has been ranked the second highest in terms of gas flaring activities and the highest in Africa having a whooping gas reserve of over 180 tcm (trillion cubic meter) that has earned it the seventh largest gas reserve worldwide (Oyedepo 2014; Shaaban and Petintin 2014). World Bank reported that Nigeria flared over 7 (seven) billion cubic meters (bcm) of gas in 2018. Nigeria produces over 2,000 billion standard cubic feet of gas yearly (Nriagu et al. 2008). This volume fluctuates regularly because of incessant unrest from militants, activities of miscreants, vandalization, leakages, and so on. Oil exploitation companies in Nigeria are unable to channel natural gas into gainful use since they consider it cost intensive to procure the facilities needed for gas capture. Gas flaring is therefore employed as a safe means to evacuate any associated natural gas during petroleum exploitation, despite its serious deleterious far reaching environmental effects. Threat of flaring gases is common in oil rigs, crude oil refineries, and plants. In advanced countries like the USA, only 1% of the associated natural gas is flared, while in Nigeria more than 60% natural gas is flared every day. This large volume of flared gas has the potential to generate substantial revenue to the government of Nigeria if properly harnessed for cooking or industrial uses (Agboola et al. 2011). The flared gas could also be used to generate (power) electricity which could end the challenge of inadequate power supply especially now that the nation is experiencing reduced oil revenues as a consequence of global decrease in oil price (Ojide et al. 2012). This will raise more revenue for Nigeria at this time when oil prices are generally low.

Nigeria losses over 1 billion naira revenue through gas flaring (Campbell 2004; Fluenta 2019) that would have contributed immensely to the nation's economy if the gas was properly captured (Buzcu-Guven and Harriss 2012). Between November 2016 and November 2017, Nigeria flared over 300 billion standard cubic feet of gas which was sold at an exchange rate of 360 naira to 1 dollar with a domestic supply rate of 1.50 USD per 1000 standard cubic feet of gas. This suggests that the nation lost a whooping sum of over 160 billion naira which could have improved her economic base (NNPC 2017). Rules and regulations governing gas flaring in Nigeria are not adequately adhered to by oil-producing companies because such laws are weakly enforced and the penalty levied on defaulters is insignificant (N10 per standard cubic feet). Oil companies in Nigeria lack modern infrastructure needed for gas capture (Okorie 2018); they perceive that such equipment is expensive, and hence its use may not be economically viable. Ironically, there has been increased preference and demand for cooking gas over other sources of fuel, yet, the colossal waste of gas flaring is still been practiced. Locally and internationally, efforts are therefore needed to ensure that regulations governing gas flaring are strictly adhered for the good of man and the environment (Otitoloju and Dan-Patrick 2010).

The composition of the flared natural gas determines the type of pollutants emitted (Fawole et al. 2016). Gas flaring emits greenhouse gases that warm the atmosphere, causing climate change (Ukala 2011); also other precursor gases, volatile organic compounds (VOCs), polyaromatic hydrocarbons (PAH), particulate matter, and black carbon from gas flares often contaminate air, soil, and water (Giwa et al. 2014). More than 200 notable toxins have been reported to emanate from gas flares including hydrogen sulfide, toluene, benzene, sulfur dioxide, nitrogen oxide, xylene, and so on (ICF 2006). Particulate matter and precursor gases are among the most toxic pollutants that affect plants and man (Giwa et al. 2014; Yaduma et al. 2013). It releases over 40 billion kilowatts of heat daily into the atmosphere (Ukala 2011). Other pollutants released during gas flaring are soot, organic carbon, particulate matter (Guttikunda and Calori 2013; Giwa et al. 2014; Zhou et al. 2014), and heavy metals (Kampa and Castanas 2008). One percent of global warming arises from gas flaring which contributes over 300 Mt of $CO_2$ into the atmosphere (IPCC 2014a; Amaechi and Biose 2016). Though the practice of gas flaring has been significantly reduced in some countries, Nigeria is yet to record a significant decline despite numerous government strategies (Ite and Ibok 2013); hence, its devastating effects are still felt (Giwa et al. 2014; Oyedepo 2014). Nigeria has lost so much money as a result of gas flaring since the commencement of oil exploitation (Odumugbo 2010). Nigeria loses well over 2 billion dollars as a result of gas flaring yearly (Campbell 2004). This income would have contributed positively to the nation's economy if the gas was properly captured (Buzcu-Guven and Harriss 2012). Gas flaring is responsible for loss of energy resources in oil companies. It affects art works, paints, and monuments (Abua and Ashua 2015; Amadi 2014; Anomohanran 2012; Donwa et al. 2015; Iyorakpo and Odibikuma 2015; Nkwocha and Pat-Mbano 2010; Olukoya 2015; Ubani and Onyejekwe 2013). It affects the general well-being of the Niger Delta dwellers, both psychologically and otherwise (Nriagu et al. 2008). Noise from flare stalks could also affect man and other living things residing around the area. This could lead to loss of some important species or even outright extinction. Pollutants emitted from gas flares are more concentrated at locations close to the flare site. Those farther away from flare sites have reduced pollutant concentrations (Ojeh 2012). Though gas-flared effects are majorly felt within 450 m radius of flare stack, factors like stack height, flare velocity, temperature, and wind speed affect its impact (Ojeh 2012). When there are notable variations in wind speed and direction, locations close to flare stack may have less concentration of pollutants, while farther locations may have more. Gases can be flared from either a high or low pressure valve, and this affects the concentration of gas emitted and also the noise generated. Global Emission Inventory stated that flaring releases three times more soot than gasoline-driven vehicles (Weyant et al. 2016). It is responsible for the disorientation of water bodies like seas and oceans, for example, for over 40 years now, the ice level of the artic sea has significantly diminished (IPCC 2013). This rise in sea level is a consequence of global warming (Bernstein et al. 2007).

The Niger Delta area of Nigeria occupies more than 7% of the nation's total land mass with not less than 20 million inhabitants (Tawari and Abowei 2012). The nation has onshore and offshore natural gas reserves located in nine states, namely, Abia, Akwa Ibom, Bayelsa, Cross River, Delta, Edo, Ondo, Imo, and Rivers. They comprise a total of 185 local government areas, housing over 800 oil-producing wells and gas operation facilities (Osuji and Onojake 2004). Niger Delta environment is constantly being polluted by activities such as gas flaring, crude oil pollution, pipeline and tanker leakages, and so on (Tawari and Abowei 2012). All these pollutants have over the period altered the ecosystem balance, crop development, and human health. They have also been found to affect microbial activities in the area (Sonibare et al. 2010). A greater percentage of rural dwellers in the Niger Delta engages in different agricultural activities especially fishing and farming as means of livelihood. Unfortunately, gas flaring has negatively affected their survival since it has negatively impacted soil, water, and the environment at large. Gas-flared areas have an elevated ambient temperature as a result of the huge amount of heat (above 45 billion kilowatts) that is released daily from gas-flared points (Aaron 2006). Crops cultivated 2 km away from gas-flared points are usually scorched by the heat emitted from the gas flares (Anomohanran 2012). Particulates from gas flares can spread as far as $2.61 \times 10^6$ km away from flare points (Ede et al. 2011). A study conducted by Gobo et al. (2009) revealed that these pollutants could trigger onset of eye, skin, and respiratory diseases. Prolonged exposure to gas flares has been reported to negatively affect blood hematological parameters (Adienbo and Nwafor 2010). Ovuakporaye et al. (2012) also reported presence of skin and respiratory diseases in residents of gas-flared areas as a result of long-term inhalation of gas-flared fumes. Gaseous pollutants from gas flares such as nitrogen and sulfur oxides are responsible for acid rain. Acid rain corrodes roofing sheets and walls of buildings. Smog that results from the reaction of nitrogen oxides and hydrocarbons in the presence of UV radiation creates poor visibility (Ana et al. 2009). Other diseases reported to be prevalent in gas-flare areas include respiratory disease, eye and skin disorders, skin cancers (Ana et al. 2009), lung cancer (Ana 2011), and so on. Between 2015 and 1965, a period of 51 years, it was reported that an approximate gas quantity of over 900 bcm and 17 bcm every year with carbon monoxide being the most abundant gas was emitted (Giwa et al. 2019). Elevated levels of deleterious heavy metals have been observed in the air of the Niger Delta area. For instance, Olobaniyi and Efe (2007) reported 0.56 mg/l of lead in the area. Gas flaring has been reported to reduce yield of crops (Dung et al. 2008; Odjugo 2007; Odjugo and Osemwenkhae 2009; Olisemauche and Avwerosuoghene 2015), induce soil pollution evidenced by increased PAH deposits in soils (Sojinu et al. 2010), and be responsible for various health defects (Bhatia and Wernham 2009; Ekpoh and Obia 2010). The people of the Niger Delta perceive that gas flaring affects their health status and disturbs ecosystem balance generally (Edino et al. 2009).

Living things rely majorly on plants for their survival. The much needed energy and nutrients required by human body for its biochemical processes can be majorly derived from plants. Herbivorous animals (goat, sheep, grass cutters) and omnivorous animals (rat, squirrel, mice) depend on plants for their survival. Plants also provide shelter in the form of vegetation cover for animals that live on land. Cassava and oil palm are among common plants that are often found near gas stacks. The growth (Ozabor and Obisesan 2015) and yield (Achi 2003) of these crops are often affected by gas flaring. Over 70% of the Niger Delta dwellers are convinced that gas flaring has negatively affected their agricultural activities (Adewale and Mustapha 2015), hence their income source. Data obtained from a study in an oil community, Ebedei in Delta, has shown that gas flaring is affecting production of tuber crops such as yam, potato, and cassava including plantain and okra as well (Ozabor and Obisesan 2015). Acid rain arising from gas-flared pollutants is poisonous to plants. It could even lead to death of plants and hence loss of vegetation cover (Amadi 2014). This could further expose the soil to erosion. Symptoms expressed by plants when attacked by acid rain include shedding of leaves, yellow leaves, loss of photosynthetic ability, and even early death (Efe 2011). Acid rain has a more deleterious effect on plants and vegetation that are closer to gas-flared system. Efe (2011) documented that acid rain has negative impact on growth and development of common staple food crops and even cash crops such as rubber. The soil has not been spared from the impacts of acid rain as it has led to loss of important nutrients needed for plant survival. Other researchers including Jacobson (1984), Neufeld et al. (1985), and Efe (2010) have all documented the effect of acid rain on different plant species; and water quality (Ogunkoy and Efi 2003; Efe and Mogborukor 2008). Plants are important sources of medicine because they possess phytochemicals and other bioactive compounds that are of great pharmaceutical importance (Epidi et al. 2016a, b). The medicinal potential of a plant is a function of its bioactive composition. In another study, Anacletus et al. (2014) and Ifemeje (2015) reported changes in nutritional and phytochemical content of some common edible vegetables such as fluted pumpkin, $H_2O$ leaf, scent leaf, and bitter leaf. In a similar study, Ujowundu et al. (2013) reported changes in phytochemical (phytate, tannin, alkaloid, and cyanogenic glycoside), proximate composition (ash, moisture, carbohydrate, and proteins), micronutrients (Ca, Na, Mg, K, and P), and vitamins (A, C, and E) in African breadfruit and Bambara groundnuts cultivated near flare sites. Phytochemical (flavonoid, tannin, alkaloid, saponin) and trace metals (Pb, Fe, Cd, Zn) of fluted pumpkin could be altered by gas flaring. Okeke and Okpala (2014) reported decreased soil nutrient in two gas-flared communities (Eket and Izombe) of the Niger Delta area.

Since the era of industrial revolution, man has been faced with the challenge of handling the problems arising from excessive crude oil exploitation, gas flaring, and climate change. Methane and carbon dioxide are the most notable greenhouse gases that are responsible for global warming and climate change. Methane has been found to be the most toxic greenhouse gas because it has the ability to induce global warming (86 times) more than $CO_2$ (IPCC 2013). To reverse global warming temperature to below 2 °C obtained during the pre-industrialization era by the year 2050, there is need to consciously reduce the volume of crude oil exploited. A certain percentage of fossil fuel, for example, about 50% natural gas with 30% crude and 80% coal, needs to be consciously left unexploited in the earth's crust to avoid undue changes in the earth's environment (McGlade and Ekins 2015; UNFCCC 2015; Zhao and Alexandroff 2019). At the 48th Intergovernmental Panel on Climate Changes conference, following close evaluation of climate change data, scientists came up with a global warming temperature target of 1.5 °C by the 2035 (IPCC 2018). It is therefore imperative to seek urgent ways of reducing greenhouse gas emission as well as seeking other sources of energy for man (Rhodes 2019). The World Bank Global Gas Flaring Reduction Initiative (GGFR 2015) has launched several international and national programs including "Zero Routine Flaring by 2030" Initiative to reduce gas flaring rates. Other bodies that launched campaigns for reduction in gas flaring include the 1992 United Nation Convention on climate change and 1997 Kyoto protocol (Malumfashi 2007). Despite all these strategies, significant changes are yet to be recorded (Elvidge et al. 2009; Fawole et al. 2016), and climatic risk is becoming more pronounced.

Climate change has gained important recognition because of its increased risks and widespread effects. It has become an issue of concern, discussion, and debate in various news media of the world. Its effect on the agricultural sector seems to be at its peak. Farmers are more aware of its existence and risks. It has affected their livelihood and general well-being, thus making adaptation measures very imperative. Climate change is a global environmental threat that consciously or unconsciously affects living and non-living things (Patz et al. 2007). Climate change is an alteration in the state of the climate over a period of 10–30 years or more, depending on man-made activities. The United Nations Framework Convention on Climate Change (UNFCCC) stated that climate change is greatly dependent on human activities that alter the atmospheric equilibrium as well as natural climate change overtime (FAO 2008). Climate change is a global phenomenon that has affected man and his environment (IPCC 2014a, b). Climatic changes manifest in the form of weather variations in temperature and precipitation. Adaptation and mitigation strategies are needed in tackling climate change impacts. Climate adaptation has to do with the capacity to manage variations in climate to minimize possible risks, explore new opportunities, or even both. Most African countries like Nigeria are more susceptible to climate change challenges because they are less aware, suffer more exposure, and have reduced potential for adaptation. Changes in climate are greatly affecting agriculture because it is a substantial source of revenue for the nation. Low level of education, poverty, insufficient infrastructure, and poor planning has contributed to increased exposure to risks in climate. It is therefore said to be involuntary in action and not limited to any particular geographical location. It can increase the virulence of disease-causing organisms (pathogens and pests) and also introduce new alien diseases arising from man-made activities or nature. Climate change could be a short or prolonged change in weather parameters that are already being experienced or forecasted as induced by anthropogenic emissions of noxious gases such as carbon dioxide (Parry et al. 2007). Climate change effects are already being felt globally and even in Africa (Creech et al. 2014). There are variations in climate change impacts from one location to another; for example, in some African countries, there have been cases of increased rainfall pattern and drought (Creech et al. 2014). These variations affect weather forecasting as there are hardly records of regularity in pattern of climatic parameters (Keller 2009). Variations in climatic factors such as increase in temperature and $CO_2$ affect plant's nutritional quality and its phytochemical composition and invariably the animals that consume such plants. It could also lead to increase incidence and severity of epidemics (EU 2015).

Farmers, their households, and stakeholders in agriculture are at greater risk of climatic change since their activities are rain dependent (Pearce et al. 1996). Agriculture is a major source of employment for Niger Delta dwellers. Adaptation to climatic vulnerability is now a major challenge to farmers, researchers, and policy makers. Most adaptation strategies are geared toward eradication of poverty among households (Mertz et al. 2009). Most farmers are already using their local and traditional practices in trying to adapt to the changes in climate with its associated risks. Climate change adaptation refers to all the steps put in place over many decades to enable a system cope with perceived or actual challenges imposed by climate change (Fankhauser 2017). There is need to put up remediative measures to help recover the ecosystem from prolonged environmental pollution and eradicate associated diseases. To enable residents of Niger Delta remain in their localities and retain their occupation as farmers, it is necessary to seek good and durable adaptation strategies. Many adaptation practices have been employed in most communities. A strategy that will likely be of immense importance is one that will improve the soil productivity of the gas-flared areas. For adaptation to be effective, the reality of climate change impacts must be appreciated (Deressa et al. 2011). The way and manner farmers view climate change determines their approaches toward handling its associated challenges. Their behavioral changes to climate change will influence adaptation options and its effects (Adger et al. 2009). Therefore, a close assessment of farmer's perception of climate risks and their specific adaptation measures will foster better understanding of their local exposure to climate risks, the farmer's adaptive capacity to cope with climate change, as well as to enhance policy formulation to tackle challenges that climate change pose on farmers. Intergovernmental Panel on Climate Change (IPCC) has postulated that the adverse effects of climate change will impact many lives globally in the years to come. Trends in rainfall are expected to drift from normal to abnormal with some areas having less rainfall while others have excess. Maturation time of crops will also be affected by increased temperatures especially in the tropics in the next 10 years. This will ultimately affect global food security and health of man (IPCC 2014a).

As a way to make agricultural activities sustainable in the Niger Delta in the face of climate risks, scientists have attempted different procedures with specific interest in those with little or no side effects to the ecosystem. Biochar is a good example of such material that has been used for soil enrichment with great potentials for soil remediation. Biochar is a biomolecule that is used to amend soils with a view to improving its biological and physical characteristics. It is made by pyrolysis of biomass at temperature range of 204–482 °C in the presence of little or no oxygen (Swanson 2013). Biochar is rich in carbon (Lehmann and Joseph 2009). It has the immense advantage of being of biological origin. Biochar improves soil physical and chemical properties, fertility, and nutrient availability (Houben et al. 2013). It reduces bioavailability of heavy metals to plants and organisms by adsorbing them (Al-Wabel et al. 2014). It also stabilizes carbon in the soil, reduces the carbon dioxide in the atmosphere (Fang et al. 2015), and reduces greenhouse gas emission (Galloway et al. 2008). Its advantage over other sources of carbon is that it keeps carbon long in the soil (Nguyen et al. 2008). Biochar is more durable and effective compared to other carbon sources; it lasts for up to 10,000 years when applied to soil. Stability of biochar depends on its pyrolysis temperature, source material, and soil type (Lehmann and Joseph 2009). Biochar helps in the breakdown of polyaromatic hydrocarbon and adsorbs heavy metals, making them biologically unavailable (Gorovtsov et al. 2018). When biochar is used as a carbon source in soils, the quantity of greenhouse gases emitted decreases by 12% yearly (Woolf et al. 2010). It is a good remediation material that reduces water loss and improves nutrients of soil (Woolf et al. 2010). Use of biochar for agricultural purposes has proven to be a good climate change adaptation practice in the face of threating global warming and climate change (Lehmann et al. 2006; Woolf et al. 2010). Biochar use is not only is environmentally friendly but also alkalinizes soil (O'Neill et al. 2009) and improves crop growth and development (Graber et al. 2010). Biochar improves soil fertility and has great potential for restructuring the ecosystem. It also increases nutrient bioavailability (Wang et al. 2012), leading to production of healthier crops as well as growth of important microorganisms (Bailey et al. 2011; Smith et al. 2010). It improves soil physicochemical properties like organic carbon, cation exchange capacity, and pH (Lehmann 2007). Under normal conditions, carbon dioxide is released from decaying plant matter and serves to balance the carbon cycle. The mechanism of action of biochar is that it forms a strong bond with carbon making it assume a very stable configuration and hence slows the process of decomposition. Biochar use traps carbon in the soil while hindering its migration into the atmosphere in the form of $CO_2$. This is therefore a good adaptation process for farmers in gas flaring areas of the Niger Delta region. This will reduce soil water loss and increase soil carbon concentrations and food security (FAO 2008). Biochar has the ability to adsorb toxins, making it unavailable for plant uptake (Sohi et al. 2009). Akachukwu et al. (2018) reported that biochar application enhanced the mineral content of *Telfairia occidentalis* that was cultivated on gas-flared polluted soil. This chapter compared and evaluated the farmers' perception of gas flaring and effect of biochar on heavy metal composition of *Telfairia occidentalis* leaves cultivated on gas-flared polluted soils and non-gas-flared soils.

## Materials and Methods

### Study Area

This research was conducted with soils collected from two locations, gas-flared area in Ohaji/Egbema L.G.A of Imo State (longitude N 05° 33.5′ and latitude E 06° 45.2′) and non-gas-flared area of Ikwuano L.G.A in Abia state (Longitude N 05° 28.5′ and Latitude E 007° 32.5′) at a depth of 0–20 cm using a sterile auger. The soil was stored in clean jut bags and transported to the greenhouse of the National Root Crops Research Institute, Umudike, Abia State. All the soil samples were collected in April.

## Sociodemographic Data

Well-structured questionnaires were administered randomly to vegetable farmers in the two study sites to assess their perception of gas flaring and its effects on vegetable cultivation and farmers' livelihood.

## Biochar Preparation

Biochar was prepared from sun-dried palm bunches by pyrolysis at 450 °C in a drum kiln. After cooling, it was milled to finer particles, sieved with a 3 $mm^2$ mesh, and subsequently stored in a clean dry bag until it was ready for use (Karamesouti and Gasparatos 2008).

## Soil Preparation and Biochar Application

Four kilograms of the soil was weighed into clean plastic container. Biochar was applied to the soils at a rate of 0 t $ha^{-1}$, 7.1 t $ha^{-1}$, 13.9 t $ha^{-1}$, 20.9 t $ha^{-1}$, and 28.0 t $ha^{-1}$ in three replicates. The soils and biochar were properly mixed, watered, and allowed to mineralize for 2 weeks.

## Procurement, Cultivation, and Growth Indices of *Telfairia occidentalis*

Mature and viable seeds of *T. occidentalis* were obtained from a local market at Ndoro, Ikwuano L.G.A, Abia state, Nigeria. Two seeds were cultivated per container and watered once every 2 days. The growth parameters were measured at an interval of 2 weeks. Matured leaves were harvested after 8 weeks of germination, and samples were air-dried and used for heavy metal determination. Growth indices were measured at 2 weeks interval for 8 weeks. Plant height was measured with the aid of a ruler. Stem diameter was measured with the aid of a vernier caliper, while number of leaves was gotten by counting.

## Heavy Metal Determination

A quantity of 0.2 mg of leaf sample was weighed into dry digestion tubes. Five milliliters of nitric acid was added, swirled, and allowed to stand overnight. Tubes were placed into a digestion block with the temperature gradually increasing from room temperature to 120 °C over about 2 h with periodical swirling of each tube. Thereafter, the temperature was increased to 180 °C until about 0.5 $cm^3$ of liquid remained. The digestion tubes were removed from the block and cooled at room temperature. The digest was diluted with ultrapure water, homogenized with a vortex mixer, and allowed to stand for a few hours prior to analysis. The heavy metal concentration was determined using an inductively coupled plasma optical emission spectrometer (ICP-OES) (Optima 8000, Perkin Elmer).

## Data Analysis

The data obtained was analyzed using the Statistical Package for Social Sciences (SPSS) version 21.0 for Windows. Mean comparison was done using one way analysis of variance (ANOVA). Duncan multiple test was used to separate means. Significant values were set at $p \leq 0.05$. Data is presented as mean ± standard deviation (SD).

## Results and Discussion

Figure 1 shows the result of effect of gas flaring on vegetable cultivation and farmers' livelihood. In all the study sites, more than 60% of vegetable farmers were female while approximately 40% were male. In the gas-flared area (test), almost all the farmers (97%) had knowledge of gas flaring while 62% only knew about gas flaring in the control. This could be because of the absence of gas flaring in

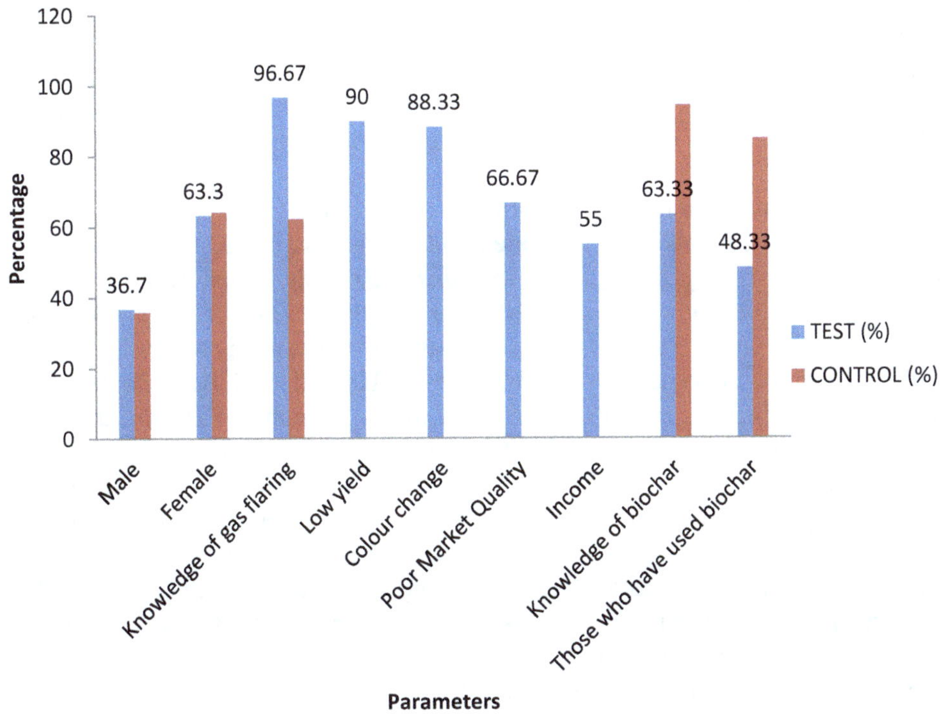


**Fig. 1** Effect of gas flaring on vegetable cultivation and farmers' livelihood

the control area. Also, farmers in the test area had knowledge of biochar (94%) and had also used it (85%) unlike the control area where fewer farmers (63%) knew about biochar and had used it (48%). More than 50% of the farmers from the test area had experienced low yield, color change, poor market quality, and income reduction as a result of gas flaring activities. This could have impacted negatively on the farmers' standard of living, causing discouragement and lack of interest in farming and invariably affect food security. It could also be a subtle cause of youth restiveness that is prevalent in the Niger Delta area. There were no bar charts for control area possibly because of the absence of gas flaring in the control area.

Table 1 shows the growth indices of *T. occidentalis* grown on the gas-flared soil and non-gas-flared soil for 8 weeks. Plant height is an important physiological property useful for assessing the growth and cell differentiation in plants (Silva et al. 2019). It is also an expression of how well a plant is able to optimally utilize soil available nutrients. There was general increase in plant height as the weeks progressed from 2 to 8 weeks for vegetables grown on GFTO and NGFTO, respectively. However, the vegetables grown on the gas-flared soils (GFTO, test) showed higher plant height than those grown on the non-gas-flared soil (NGFTO, control). The 0 t $ha^{-1}$ had no evidence of growth after the first 2 weeks. The 7.1 t $ha^{-1}$ group had the highest initial plant height (25.77) for the test vegetable, while the 28.0 t $ha^{-1}$ group had the highest (25.87) for the control. The pattern of plant height obtained is similar to that of maize plant treated with *Daniella oliviera* wood biochar reported by Aminu and Shamsuddeen (2019). Stem diameter of test vegetables was higher than the control in a non-dose-dependent manner and also reduced with increase in number of weeks in both the test and control vegetables; however, their differences were not statistically different. Dispenza et al. (2016) obtained a dose-dependent increase in stem diameter for substrates of *Euphorbicia* treated with 20%, 60%, and 80% conifer wood biochar. Number of leaves increased with increase in weeks (Okonwu and Mensah 2012).

Table 2 presents the heavy concentration of *T. occidentalis* leaves cultivated on biochar-remediated soils from gas-flared and non-gas-flared area. The Fe and Zn concentrations of the gas-flared *T. occidentalis* (GFTO) decreased from 13.9 t $ha^{-1}$ to 28.0 t $ha^{-1}$ biochar while that of the non-gas-flared *T. occidentalis* (NGFTO) decreased from 0 t $ha^{-1}$ to 20.9 t $ha^{-1}$ biochar. There was no significant ($p > 0.05$) difference in Fe concentration between the GFTO and NGFTO except for the 13.9 t $ha^{-1}$ biochar group that was markedly increased in the GFTO vegetable. No significant difference was observed in Cu concentration for GFTO and NGFTO for all the groups except for the 7.1 t $ha^{-1}$, 13.9 t $ha^{-1}$, and 20.9 t $ha^{-1}$ biochar groups that were significantly lower for NGFTO compared to the control. Lead (Pb) and Cr concentrations were significantly ($p < 0.05$) higher for GFTO from 7.1 t $ha^{-1}$ biochar, while its concentration decreased from 7.1 t $ha^{-1}$ to 20.9 t $ha^{-1}$ biochar compared to their control. Cobalt concentration was highest for GFTO treated with 7.1 t $ha^{-1}$ biochar while that of 28.0 t $ha^{-1}$ GFTO had the least. In the NGFTO, the cobalt concentration decreased from 7.1 t $ha^{-1}$ to 20.9 t $ha^{-1}$ biochar. Cadmium concentration in both the GFTO and NGFTO decreased with increasing biochar concentration. Nickel concentration was highest in the GFTO-treated with 7.1 t $ha^{-1}$

**Table 1** Effect of biochar remediation on the growth parameters of *T. occidentalis* cultivated on gas-flared and non-gas-flared soils

| Parameter | Biochar rate (t $ha^{-1}$) | 2 weeks | | 4 weeks | | 6 weeks | | 8 weeks | |
|---|---|---|---|---|---|---|---|---|---|
| | | GFTO | NGFTO | GFTO | NGFTO | GFTO | NGFTO | GFTO | NGFTO |
| **Plant height (cm)** | 0 | $18.50 \pm 6.01^{bc*}$ | $0^{d}$ | $34.57 \pm 3.44^{c}$ | $27.97 \pm 4.24^{c}$ | $34.5 \pm 2.29^{c}*$ | $47.9 \pm 4.85$ | $36.63 \pm 0.82^{b}$ | $60.77 \pm 6.20^{a}$ |
| | 7.1 | $25.77 \pm 3.39^{a*}$ | $9.67 \pm 0.577^{c}$ | $45.4 \pm 3.77^{b}*$ | $25.07 \pm 0.81^{c}$ | $62.60 \pm 2.59^{a}*$ | $41.17 \pm 5.53^{a}$ | $68.4 \pm 5.11^{a}$ | $58.50 \pm 3.91^{a}$ |
| | 13.9 | $22.73 \pm 0.92^{ab}$ | $22.00 \pm 2.00^{b}$ | $48.97 \pm 3.11^{b}$ | $38.00 \pm 5.63^{b}$ | $44.07 \pm 4.15^{b}$ | $43.57 \pm 4.29^{a}$ | $49.35 \pm 4.65^{b}$ | $49.17 \pm 3.06^{b}$ |
| | 20.9 | $22.17 \pm 3.18^{ab}$ | $21.3 \pm 1.21^{b}$ | $56.97 \pm 2.97^{a}*$ | $42.53 \pm 2.41^{b}$ | $63.05 \pm 1.50^{a}*$ | $50.83 \pm 5.97^{a}$ | $68.17 \pm 4.16^{a}$ | $57.43 \pm 6.19^{a}$ |
| | 28.0 | $16.00 \pm 0.35^{c}$ | $25.87 \pm 3.32^{a}$ | $56.03 \pm 2.44^{aa}$ | $4b9.00 \pm 1.32^{a}$ | $63.97 \pm 5.35^{a}$ | $50.33 \pm 6.43^{a}$ | $75.3 \pm 15.89^{a}$ | $63.50 \pm 1.32^{a}$ |
| **Stem diameter (cm)** | 0 | $4.81 \pm 0.93^{a}*$ | $0^{d}$ | $3.54 \pm 0.40^{a}$ | $3.51 \pm 0.49^{ab}$ | $3.44 \pm 0.86^{a}*$ | $2.79 \pm 0.25^{c}$ | $3.06 \pm 0.47^{a}*$ | $2.71 \pm 0.19^{a}$ |
| | 7.1 | $4.72 \pm 0.61^{a}*$ | $3.84 \pm 0.035^{c}$ | $3.48 \pm 0.22^{a}*$ | $2.64 \pm 0.75^{b}$ | $3.18 \pm 0.36^{a}*$ | $2.90 \pm 0.34^{bc}$ | $3.37 \pm 0.42^{a}*$ | $3.03 \pm 0.08^{a}$ |
| | 13.9 | $5.13 \pm 0.36^{a}$ | $4.63 \pm 0.01^{ab}$ | $3.76 \pm 0.26^{a}*$ | $2.59 \pm 0.69^{b}$ | $3.12 \pm 1.17^{a}*$ | $3.64 \pm 0.35^{a}$ | $3.28 \pm 0.53^{a}*$ | $3.13 \pm 0.54^{a}$ |
| | 20.9 | $4.75 \pm 0.69^{a}*$ | $4.92 \pm 0.40^{a}$ | $3.87 \pm 0.85^{a}*$ | $4.31 \pm 1.09^{a}$ | $4.03 \pm 0.91^{a}*$ | $3.43 \pm 0.24^{ab}$ | $3.79 \pm 1.15^{a}*$ | $2.99 \pm 0.54^{a}$ |
| | 28.0 | $4.56 \pm 0.63^{a}*$ | $4.19 \pm 0.41^{bc}$ | $3.66 \pm 0.92^{a}*$ | $3.8 \pm 0.55^{ab}$ | $3.63 \pm 0.53^{a}*$ | $3.55 \pm 0.42^{a}$ | $3.30 \pm 0.21^{a}*$ | $3.60 \pm 0.64^{a}$ |
| **Number of leaves (cm)** | 0 | $4.00 \pm 0.00^{a}$ | $0^{d}$ | $5.67 \pm 1.53^{b}$ | $5.00 \pm 1.73^{a}$ | $7.67 \pm 2.08^{a}$ | $8.67 \pm 0.58^{bc}$ | $9.67 \pm 1.15^{a}$ | $12.00 \pm 1.00^{ab}$ |
| | 7.1 | $4.00 \pm 0.71^{a}*$ | $1.00 \pm .000^{c}$ | $8.67 \pm 2.08^{a}$ | $5.33 \pm 1.53^{a}$ | $10.00 \pm 2.00^{a}$ | $9.00 \pm 1.00^{abc}$ | $10.67 \pm 2.89^{a}$ | $10.33 \pm 1.16^{ab}$ |
| | 13.9 | $3.00 \pm 1.41^{a}$ | $2.67 \pm 0.578^{b}$ | $6.00 \pm 1.00^{ab}$ | $6.67 \pm 1.15^{a}$ | $8.33 \pm 2.08^{a}$ | $8.00 \pm 1.73^{c}$ | $9.50 \pm 1.50^{a}$ | $8.67 \pm 2.52^{b}$ |
| | 20.9 | $4.00 \pm 0.71^{a}$ | $3.67 \pm 0.57^{a}$ | $8.33 \pm 1.15^{ab}$ | $7.33 \pm 1.15^{a}$ | $8.67 \pm 2.08^{a}$ | $10.33 \pm 0.58^{ab}$ | $12.00 \pm 2.00^{a}$ | $12.67 \pm 2.52^{a}$ |
| | 28.0 | $3.00 \pm 1.41^{a}$ | $3.67 \pm 0.57^{a}$ | $7.33 \pm 1.15^{ab}$ | $7.67 \pm 1.53^{a}$ | $9.33 \pm 1.15^{a}$ | $10.67 \pm 0.58^{a}$ | $12.33 \pm 1.15^{a}$ | $12.00 \pm 1.00^{ab}$ |

Values are mean ± standard deviation of triplicate determinations.
Values with different superscript on the same column are significantly different ($p < 0.05$) while values marked asterick (*) are significantly different from their controls.

**Table 2** Effect of biochar remediation on the heavy metal concentration of *T. occidentalis* cultivated on gas-flared and non-gas-flared soils

| Heavy metal (mg kg$^{-1}$) | GP 0 (0 t ha$^{-1}$) | GP I (7.1 t ha$^{-1}$) | GP II (13.9 t ha$^{-1}$) | GP III (20.9 t ha$^{-1}$) | GP IV (28.0 t ha$^{-1}$) |
|---|---|---|---|---|---|
| Fe (GFTO) | $103.25 \pm 9.73$ | $108.57 \pm 12.03^{a}$ | $135.80 \pm 20.06^{a*}$ | $97.74 \pm 27.26^{a}$ | $93.48 \pm 9.73^{a}$ |
| Fe (NGFTO) | $126.14 \pm 6.56^{a}$ | $98.15 \pm 2.51^{bc}$ | $97.79 \pm 3.31^{bc}$ | $89.24 \pm 3.73^{c}$ | $117.56 \pm 10.50^{ab}$ |
| Cu (GFTO) | $0.55 \pm 0.05$ | $0.58 \pm 0.06^{a}$ | $0.72 \pm 0.11^{a*}$ | $0.52 \pm 0.14^{a}$ | $0.50 \pm 0.50^{a}$ |
| Cu (NGFTO) | $0.67 \pm 0.03^{a}$ | $0.52 \pm 0.01^{bc}$ | $0.52 \pm 0.09^{bc}$ | $0.47 \pm 0.02^{c}$ | $0.62 \pm 0.06^{ab}$ |
| Zn (GFTO) | $20.65 \pm 1.95$ | $21.71 \pm 2.41^{a}$ | $27.16 \pm 4.01^{a*}$ | $19.55 \pm 5.45^{a}$ | $18.70 \pm 1.95^{a}$ |
| Zn (NGFTO) | $25.23 \pm 1.31^{a}$ | $19.63 \pm 0.50^{bc}$ | $19.56 \pm 3.49^{bc}$ | $17.85 \pm 0.75$ | $23.51 \pm 2.10^{ab}$ |
| Pb (GFTO) | $0.26 \pm 0.02^{bc}$ | $0.37 \pm 0.01^{a}$ | $0.30 \pm 0.01^{bc}$ | $0.32 \pm 0.03^{ab}$ | $0.24 \pm 0.02^{c}$ |
| Pb (NGFTO) | $0.32 \pm 0.02^{a}$ | $0.25 \pm 0.01^{bc}$ | $0.25 \pm 0.04^{bc}$ | $0.23 \pm 0.01^{c}$ | $0.30 \pm 0.03^{ab}$ |
| Cr (GFTO) | $2.66 \pm 0.25^{bc}$ | $3.77 \pm 0.14^{a}$ | $2.99 \pm 0.09^{bc}$ | $3.23 \pm 0.30^{ab}$ | $2.41 \pm 0.25^{c}$ |
| Cr (NGFTO) | $3.25 \pm 0.17^{a}$ | $2.53 \pm 0.06^{bc}$ | $2.52 \pm 0.04^{bc}$ | $2.30 \pm 0.10$ | $3.03 \pm 0.27^{ab}$ |
| Co (GFTO) | $6.51 \pm 0.61^{bc}$ | $9.21 \pm 0.34$ | $7.31 \pm 0.21$ | $7.88 \pm 0.72^{ab}$ | $5.89 \pm 0.61$ |
| Co (NGFTO) | $7.95 \pm 0.41^{a}$ | $6.18 \pm 0.16$ | $6.16 \pm 1.10$ | $5.62 \pm 0.24$ | $7.41 \pm 0.66$ |
| Cd (GFTO) | $0.06 \pm 0.01^{c}$ | $0.08 \pm 0.00^{a}$ | $0.07 \pm 0.00^{bc}$ | $0.07 \pm 0.01^{ab}$ | $0.05 \pm 0.01^{c}$ |
| Cd (NGFTO) | $0.07 \pm 0.00^{a}$ | $0.06 \pm 0.00^{bc}$ | $0.06 \pm 0.01^{bc}$ | $0.05 \pm 0.00$ | $0.07 \pm 0.01^{ab}$ |
| Ni (GFTO) | $3.08 \pm 0.29$ | $4.36 \pm 0.16^{a}$ | $3.46 \pm 0.10^{bc}$ | $3.73 \pm 0.34^{ab}$ | $2.79 \pm 0.29$ |
| Ni (NGFTO) | $3.77 \pm 0.20^{a}$ | $2.93 \pm 0.07^{bc}$ | $2.92 \pm 0.52^{bc}$ | $2.66 \pm 0.11$ | $3.51 \pm 0.31^{ab}$ |

Values are mean ± standard deviation of triplicate determinations.

Values with different superscript (a, b, c) on the same row are significantly different ($p < 0.05$) while values marked asterick (*) are significantly different from their controls.

GFTO, Gas-flared *Telfairia occidentalis*; NGFTO, Non-gas-flared *Telfairia occidentalis* biochar and lowest for the 28.0 t $ha^{-1}$ biochar-treated group, while the NGFTO treated with 20.9 t $ha^{-1}$ biochar had the least nickel concentration. Biochar administered to the two soils reduced uptake by plants as shown by our result. Human beings are often exposed to heavy metal contamination through the food chain as they are transferred from soil to plants (Khan et al. 2010). They cause serious health challenges when their concentrations exceed the normal threshold (Al-Wabel et al. 2014). Some heavy metals are known to alter normal body metabolism, disrupt transfer of hereditary materials from parents to offsprings, and affect proper growth and development of fetus (Ali et al. 2013). Heavy metals such as chromium, copper, cadmium, and lead are well known for their toxicity when consumed at concentrations beyond the permissible limits (Dursun 2006; Kurniawan et al. 2006). Copper contamination can cause gastrointestinal cancer in humans (Turkdogan et al. 2003). Cadmium toxicity is responsible for kidney damage and can cause "itai-itai" in man. Zinc is an essential element that is toxic at higher concentrations (Baccio et al. 2005). Chromium is harmful to plants and animals; chromium (IV) ion alters soil biological activities and is carcinogenic (Javied et al. 2009). Cobalt causes bone marrow hyperplasia, acute poisoning, allergic reactions, seizures, and paralysis of nervous system. Lead is very poisonous; it affects the cardiovascular system and causes stroke and cognitive impairment (Evangelou et al. 2007). Our finding has shown that heavy metal adsorption by the *T. occidentalis* leaves decreased as the rate of biochar application increased with 20.9 t $ha^{-1}$ being the most effective. This result suggest that using biochar to cultivate this vegetable could lead to production of better quality vegetables, reducing the prevalence of heavy metal toxicity and hence improving nutrition of the consumers.

## Conclusion

The chapter has shown that vegetable farming is mostly engaged by females. Biochar use enhanced the growth parameters and reduced heavy metal uptake by the plants. Therefore, biochar use should be encouraged to remediate heavy metal polluted soils, to ensure maximum crop yield and food security. Increased sensitization is needed to encourage the use of biochar in gas flaring polluted areas to ensure that good quality vegetables are available for consumers.

**Acknowledgments** This research is supported by funding from the Department for International Development (DFID) under the Climate Impacts Research and Leadership Enhancement (CIRCLE) program.

## References

Aaron KK (2006) Human rights violation and environmental degradation in the Niger-Delta. In: Porter E, Offord B (eds) Activating human rights. Barnes, Oxford/New York

Abua MA, Ashua SW (2015) The impact of gas flaring on plant diversity in Ibeno local government area. J Agric Ecol Res Int 4(1):10–17

Achi C (2003) Hydrocarbon exploitation, environmental degradation and poverty: the Niger Delta experience. In: Diffuse pollution conference, Dublin 2003 2B policy/socio-economics

Adewale OO, Mustapha U (2015) The impact of gas flaring in Nigeria. Int J Sci Technol Soc 3 (2):40–50

Adger WN et al (2009) Are there social limits to adaptation to climate change? Clim Chang 93 (3–4):335–354

Adienbo OM, Nwafor A (2010) Effect of prolong exposure to gas flaring on some haematological parameters of humans in the Niger Delta region of Nigeria. J Appl Sci Environ Manag 14 (1):13–15

Agboola MO, Nwulu NI, Egelioglu F, Agboola OP (2011) Gas flaring in Nigeria: opportunity for household cooking utilization. Int J Therm Environ Eng 2:73

Ajugwo AO (2013) Negative effects of gas flaring: the Nigerian experience. J Environ Pollut Hum Health 1:6–8

Akachukwu D, Gbadegesin MA, Ojimelukwe PC, Atkinson CJ (2018) Biochar remediation improves the leaf mineral composition of *Telfairia occidentalis* grown on gas flared soil. Plants 7(3):57

Ali H, Khan E, Sajad MA (2013) Phytoremediation of heavy metals-concepts and applications. Chemosphere 91(7):869–881

Al-Wabel MI, Usman ARA, El-Naggar AH, Aly AA, Ibrahim HM, Elmaghraby S, Al-Omran A (2014) Conocarpus biochar as a soil amendment for reducing heavy metal availability and uptake by maize plants. Saudi J Biol Sci 22:503–511

Amadi AN (2014) Impact of gas-flaring on the quality of rain water, groundwater and surface water in parts of Eastern Niger Delta, Nigeria. J Geosci Geomatics 2(3):114–119

Amaechi CF, Biose E (2016) Gas flaring: carbon dioxide contribution to global warming. J Appl Sci Environ Manag 20:309–317. https://www.ajol.info/index.php/jasem/article/view/140405. Accessed 19 Dec 2019

Aminu H, Shamsuddeen L (2019) Effect of Daniella olivera wood biochar on the growth and biomass of maize (*Zeamays* L.) in Lafia, North Central Nigeria. J Dryland Agric 5(3):25–32

Ana GR (2011) Air pollution in the Niger Delta area: scope, challenges and remedies. In: Khallaf M (ed) The impact of air pollution on health, economy, environment and agricultural sources. InTech. ISBN: 978-953-307-528-0. http://www.intechopen.com/books/the-impact-ofairpollution-on-health-economy-environment-and-agricultural-sources/airpollution-in-the-niger-delta-area-scopechallenges-and-remedies

Ana GREE, Sridhar MKC, Bamgboye EA (2009) Environmental risk factors and health outcomes in selected communities of the Niger Delta area, Nigeria. Perspect Public Health 129(4): 183–191

Anacletus F, Adisa OI, Uwakwe A (2014) Effect of gas flaring on some phytochemicals and trace metals of fluted pumpkin (*Telferia occidentalis*). J Environ Earth Sci 4(16):1–5

Anomohanran O (2012) Thermal effect of gas flaring at ebedei area of Delta State, Nigeria. Pac J Sci Technol 13:555–560. http://www.akamaiuniversity.us/PJST13_2_555.pdf. Accessed 19 Dec 2019

Baccio D, Kopriva S, Sebastiani H, Rennenberg L (2005) Does glutathione metabolism have a role in the defence of poplar against zinc excess. New Phytol 167:73–80

Bailey VL, Fansler SJ, Smith JL, Bolton H (2011) Reconciling apparent variability in effects of biochar amendment on soil enzyme activities by assay optimization. Soil Biol Biochem 43:296

Bernstein L, Bosch P, Canziani O, Chen Z, Christ R, Davidson O (2007) Climate change 2007: synthesis report. Summary for policymakers. IPCC, Geneva, p 22

Bhatia R, Wernham A (2009) Integrating human health into environmental impact assessment: an unrealized opportunity for environmental health and justice. Ciênc Saúde Colet 14(4): 1159–1175

Buzcu-Guven B, Harriss R (2012) Extent, impacts and remedies of global gas flaring and venting. Carbon Manag 3(1):95–108

Campbell LGM (2004) A voluntary standard for global gas flaring and venting reduction. World Bank Group, Oil, Gas, Mining and Chemicals Department, Washington, DC

Creech H, Akoh B, Parry JE (2014) ICTs for climate change adaptation in Africa. World Bank, Washington, DC

Deressa TT, Hassan RM, Ringler C (2011) Perception of and adaptation to climate change by farmers in the Nile basin of Ethiopia. J Agric Sci 149:23

Dispenza V, De Pasquale C, Fascella G, Mammano MM, Alonzo G (2016) Use of biochar as peat substitute for growing substrates of Euphorbia × lomi potted plants. Span J Agric Res 14(4): e0908

Donwa PA, Mgbame CO, Utomwen OA (2015) Gas flaring in the oil and gas sector in Nigeria. Int J Commer Manag Res 1(1):28–39

Dung E, Bombom L, Agusomu T (2008) The effects of gas flaring on crops in the Niger Delta, Nigeria. GeoJournal 73(4):297–305

Dursun AYA (2006) Comparative study on determination of the equilibrium, kinetic and thermodynamic parameters of biosorption of copper(II) and lead(II) ions onto pretreated *Aspergillus* Niger. Biochem Eng J 28(2):187–195

Ede PN, Edokpa DO (2015) Regional air quality of the Nigeria's Niger Delta. Open J Air Pollut 4:7–15

Ede PN, Edokpa DO, Ayodeji O (2011) Aspects of air quality status of Bonny Island, Nigeria attributed to an LNG plant. Energy Environ 22(7):891–909

Edino MO, Nsofor GN, Bombom LS (2009) Perceptions and attitudes towards gas flaring in the Niger Delta, Nigeria. Enviromentalist 30:67–75

Efe SI (2010) Spatial variation in acid and some heavy metal composition of rainwater harvesting in the oil producing region of Nigeria. Nat Hazards 55:307

Efe SI (2011) Spatial variation of acid rain and its ecological effect in Nigeria. In: Proceedings of the environmental management conference, Federal University of Agriculture, Abeokuta, pp 381–396

Efe SI, Mogborukor JOA (2008) Acid rain in the Niger Delta region: implication on water resource crises. In: Proceedings, international conference on the Nigerian state, oil industry and the Niger Delta, 11–13th March, 2008, Glory land Cultural Centre, Yenagoa

Ekpoh I, Obia A (2010) The role of gas flaring in the rapid corrosion of zinc roofs in the Niger Delta region of Nigeria. Environmentalist 30(4):347–352

Elvidge CD, Ziskin D, Baugh KE, Tuttle BT, Ghosh T, Pack DW, Erwin EH, Zhizhin M (2009) A fifteen year record of global natural gas flaring derived from satellite data. Energies 2(3): 595–622

Emam EA (2016) Environmental pollution and measurement of gas flaring. Int J Innov Res Sci Eng Technol 2:252–262. https://pdfs.semanticscholar.org/0463/5c29995aefa3f8f7d431d4ca6a35d22c4813.pdf. Accessed 19 Dec 2019

Epidi JO, Izah SC, Ohimain EI, Epidi TT (2016a) Phytochemical, antibacterial and synergistic potency of tissues of *Vitex grandifolia*. Biotechnol Res 2(2):69–76

Epidi JO, Izah SC, Ohimain EI (2016b) Antibacterial and synergistic efficacy of extracts of *Alstonia boonei* tissues. Br J Appl Res 1(1):0021–0026

European Union (2015) Life and climate change bulletin. Publications Office of the European Union, Luxembourg, pp 1–113. https://www.google.com/search?client=firefox-b-d&q=life_climate_change_adaptation_en

Evangelou MWH, Ebel M, Schaeffer A (2007) Chelate assisted phytoextraction of heavy metals from soil. Effect, mechanism, toxicity, and fate of chelating agents. Chemosphere 68 (6):989–1003

Fang Y, Singh B, Singh BP (2015) Effect of temperature on biochar priming effects and its stability in soils. Soil Biol Biochem 80:136–145

Fankhauser S (2017) Adaptation to climate change. Ann Rev Resour Econ 9(1):209–230

FAO (Food and Agriculture Organization of the United Nations) FAOSTAT (2008)

Fawole OG, Cai XM, MacKenzie AR (2016) Gas flaring and resultant air pollution: a review focusing on black carbon. Environ Pollut 216:182–197

Fluenta (2019) Reducing flaring in Nigeria: presidential approval. https://www.fluenta.com/reducing-flaring-in-nigeria-presidential-approval/. Accessed 16 July 2020

Galloway JN, Townsend AR, Erisman JW, Bekunda M, Cai Z, Freney JR, Martinelli LA, Seitzinger SP, Sutton MA (2008) Transformation of the nitrogen cycle: recent trends, questions, and potential solutions. Science 320:889–892

Giwa SO, Adama OO, Akinyemi OO (2014) Baseline black carbon emissions for gas flaring in the Niger Delta region of Nigeria. J Nat Gas Sci Eng 20:373–379

Giwa SO, Kuye SI, Adama KO (2019) Gas flaring attendant impacts of criteria and particulate pollutants: a case of Niger Delta region of Nigeria. J King Saud Univ Eng Sci 31:209–217

Global Gas Flaring Reduction Partnership (GGFR) (2015) https://www.worldbank.org/en/programs/gasflaringreduction. Accessed 5 Nov 2019

Gobo AE, Richard G, Ubong IU (2009) Health impact of gas flares on Igwurata / Umuchem communities in Rivers State. J Appl Sci Environ Manage 13(3):27–33

Gorovtsov AV, Sazykin IS, Sazykina MA (2018) The influence of heavy metals, polyaromatic hydrocarbons, and polychlorinated biphenyls pollution on the development of antibiotic resistance in soils. Environ Sci Pollut Res 25(10):9283–9292

Graber ER, Harel YM, Kolton M, Cytryn E, Silber A, David DR, Tsechansky L, Borenshtein M, Elad Y (2010) Biochar impact on development and productivity of pepper and tomato grown in fertigated soilless media. Plant Soil 337:481–496

Guttikunda SK, Calori G (2013) A GIS based emissions inventory at 1 km × 1 km spatial resolution for air pollution analysis in Delhi, India. Atmos Environ 67:101–111

Houben D, Evrard L, Sonnet P (2013) Mobility, bioavailability and PH-dependent leaching of cadmium, zinc and lead in a contaminated soil amended with biochar. Chemosphere 92:1450–1457

ICF (2006) Nigeria: guidebook for carbon credit development for flare reduction projects. International. June 2006

Ifemeje JC (2015) Effect of gas flaring on the anti-nutritional composition of four green leafy vegetables from Eleme in Rivers State, Nigeria. Int J Environ Pollut Solut 3(1):31–37. Columbia International Publishing

IPCC (2013) Climate change 2013. The physical science basis. Contribution of working group I to the fifth assessment report of the Intergovernmental Panel on Climate Change. Cambridge University Press, Cambridge, UK, 2014

IPCC (2014a) Climate change 2014. Impacts, adaptation and vulnerability-contribution of working group II to the fifth assessment report of Intergovernmental Panel on Climate Change. Cambridge University Press, Cambridge, UK

IPCC (2014b) Climate change 2014: synthesis report. Intergovernmental panel on climate change, Geneva

IPCC (2018) Global warming of 1.5 °C. An IPCC Special report on the impacts of global warming of 1.5 °C above pre-industrial levels and related global greenhouse gas emission pathways, in the context of strengthening the global response to the threat of climate change. Intergovernamental Panel on Climate Change. https://www.ipcc.ch/sr15/download/#full. Accessed 5 Nov 2019

Ismail OS, Umukoro GE (2012) Global impact of gas flaring. Energy Power Eng 4(4):290–302

Ite AE, Ibok UJ (2013) Gas flaring and venting associated with petroleum exploration and production in the Nigeria's Niger Delta. Am J Environ Prot 1:70–77

Iyorakpo J, Odibikuma PW (2015) Impact of gas flaring on the built environment: the case of Ogba/Egbema/Ndoni Local Govt Area, Rivers State, Nigeria. Eur Sci J 11(26):83–95

Jacobson JS (1984) Effect of acidic aerosol, fog, mist and rain on crops and trees. Philos Trans R Soc Lond B 49:327–338

Javied S, Mehmood T, Choodhry MM, Tufail M, Irfan N (2009) Heavy metal pollution from phosphate rock used for the production of fertilizer in Pakistan. Microchem J 91:94–99

Kampa M, Castanas E (2008) Human health effects of air pollution. Environ Pollut 151:362–367
Karamesouti M, Gasparatos D (2008) Sustainable management of soil phosphorus in a changing world. In: Rakshit S, Abhilash PC, Singh HB, Ghosh S (eds) Adaptive soil management – from theory to practices. Springer, Berlin
Keller M (2009) Climate risks and development projects: an assessment report for a community-level project in Guduru, Oromiya, Ethiopia, Swiss Interchurch Aid (HEKS)
Khan S, Rehman S, Khan AZ, Khan MA, Shah MT (2010) Soil and vegetables enrichment with heavy metals from geological sources in Gilgit, Northern Pakistan. Ecotoxicol Environ Saf 73:1820–1827
Kurniawan TA, Chan GY, Lo WH, Babel S (2006) Comparisons of low-cost adsorbents for treating waste waters laden with heavy metals. Sci Total Environ 366(2–3):409–426
Lehmann J (2007) Bio-energy in the black. Front Ecol Environ 5:381–387
Lehmann J, Joseph S (2009) Biochar for environmental management: an introduction. In: Lehmann J, Joseph S (eds) Biochar for environmental management, science and technology. Earthscan, London, pp 1–12
Lehmann J, Gaunt J, Rondon M (2006) Biochar sequestration in terrestrial ecosystems–a review. Mitig Adapt Strateg Glob Chang 11:395–419
Malumfashi GI (2007) Phase-out of gas flaring in Nigeria by 2008: the prospects of a multi-win project. Review of the regulatory, environmental and socio-economic issues. Niger Gas Flaring Pet Train J 4:1–39
McGlade C, Ekins P (2015) The geographical distribution of fossil fuels unused when limiting global warming to 2 °C. Nature 517:187–190
Mertz O, Mbow C, Reenberg A, Diouf A (2009) Farmers' perceptions of climate change and agricultural adaptation strategies in rural Sahel. Environ Manag 43(5):804–816
Mohammed AA, Mohammed MA, Ahmed MM (2016) Recovery of natural gas liquids and produce liquified petroleum gas from associated natural gas. Sudan University of Science & Technology. Avaialable at: http://repository.sustech.edu/handle/123456789/15090
Neufeld HS, Jernstedt JA, Haines BL (1985) Direct foliar effects of simulated acid rain, I. Damage, growth and gas exchange. New Phytol 99(3):389–405
Nguyen B, Lehmann J, Kinyangi J, Smernik R, Riha SJ, Engelhard MH (2008) Long-term black carbon dynamics in cultivated soil. Biogeochemistry 89:295–308
Nigerian National Petroleum Corporation (2017) NNPC: Nigeria drops to seventh highest gas flaring country. https://www.thisdaylive.com/index.php/2017/04/03/nnpc-nigeria-drops-toseventh-highest-gas-flaring-country/. Accessed 20 July 2020
Nkwocha EE, Pat-Mbano EC (2010) Effect of gas flaring on buildings in the oil producing rural communities of River State, Nigeria. Afr Res Rev 4(2):90–102
Nriagu J, Udofia EA, Ekong I, Ebuk O, Sambo AS (2008) The role of energy in achieving millennium development goal (MDGs). Keynote address at the national engineering technology conference (NETec 2008). Ahmadu Bello University, Zaria held on 1st of April, 2008
O'Neill B, Grossman J, Tsai MT, Gomes JE, Lehmann J, Peterson J, Neves E, Thies JE (2009) Bacterial community composition in Brazilian Anthrosols and adjacent soils characterized using culturing and molecular identification. Microb Ecol 58:23–35
Odjugo PAO (2007) Some effects of gas flaring on the microclimate of yam and cassava production in Erhorike and environs. Delta State, Nigeria. Niger Geogr J 5:43
Odjugo PAO, Osemwenkhae EJ (2009) Natural gas flaring affects microclimate and reduces maize (*Zea mays*) yield. Int J Agric Biol 11:408–412
Odumugbo CA (2010) Natural gas utilisation in Nigeria: challenges and opportunities. J Nat Gas Sci Eng 2:310–316
Ogunkoy OO, Efi EJ (2003) Rainfall quality and sources of rainwater acidity in Warri area of the Niger Delta, Nigeria. J Min Geol 39(2):125–130
Ohimain EI (2013a) Can the Nigerian biofuel policy and incentives (2007) transform Nigeria into a biofuel economy? Energy Policy 54:352–359

Ohimain EI (2013b) The challenges of liquid transportation fuels in Nigeria and the emergence of the Nigerian automotive biofuel programme. Res J Appl Sci Eng Technol 5(16):4058–4065

Ojeh VN (2012) Sustainable development and gas flaring activities: a case study of Ebedei area of Ukwuani LGA, Delta State, Nigeria. Resour Environ 2(4):169–174

Ojide M, Kareem S, Kari F, Gazi A, Matuin O (2012) Impact of gas industry on sustainable economy in Nigeria: further estimations through e-view. J Appl Sci 12:2244–2251

Okeke PN, Okpala CQ (2014) Effects of gas flaring on selected arable soil quality indicators in the Niger Delta, Nigeria. Sky J Soil Sci Environ Manag 3(1):001–005

Okonwu K, Mensah SI (2012) Effects of NPK (15:15:15) fertilizer on some growth indices of pumpkin. Asian J Agric Res 6:137–143

Okorie A (2018) The dominance of foreign capital and its impact on Indigenous technology development in the production of liquefied natural gas in Nigeria. Rev Afr Polit Econ 45:478–490

Olisemauche OO, Avwerosuoghene OP (2015) The effect of gas flaring on agricultural production of Okpai, Ndukwa East Local Government Area, Delta State, Nigeria. Stand Sci Res Essays 3 (9):266–272

Olobaniyi SB, Efe SI (2007) Comparative assessment of rainwater and groundwater in the oil producing area of Nigeria: environment and health implication. J Environ Health 6 (2):111–117

Olukoya OAP (2015) Negative effects of gas flaring on buildings and public health in oil producing communities: the Ogbia community, Bayelsa State case. Int J Environ Monit Prot 2(5):52–61

Orlando O (2006) Construction and networking of gas distribution infrastructure, Chartered Institute of Bankers of Nigeria, (2006) Oil and gas financing in Nigeria issues, challenges and prospects. The Chartered Institute of Bankers of Nigeria, Lagos, p 243

Osuji LC, Onojake CM (2004) Trace heavy metals associated with crude oil: a case study of Ebocha-8 oil-spill-polluted site in Niger Delta, Nigeria. Chem Biodivers 1(11):1708–1715

Otitoloju A, Dan-Patrick J (2010) Effects of gas flaring on blood parameters and respiratory system of laboratory mice, *Mus musculus*. Environmentalist 30:340–346

Ovuakporaye SI, Aloamaka CP, Ojieh AE, Ejebe DE, Mordi JC (2012) Effect of gas flaring on lung function among residents in gas flaring community in Delta State, Nigeria. Res J Environ Earth Sci 4(5):525–528

Oyedepo SO (2014) Towards achieving energy for sustainable development in Nigeria. Renew Sust Energ Rev 34:255–272

Ozabor F, Obisesan A (2015) Gas flaring: impacts on temperature, agriculture and the people of Ebedei in Delta State Nigeria. J Sustain Soc 4(2):5–12

Parry ML, Canziani OF, Palutikof JP, van der Linden PJ, Hanson CE (eds) (2007) Cross-chapter case study. In: Climate change 2007: impacts, adaptation and vulnerability. Contribution of working group II to the fourth assessment report of the Intergovernmental Panel on Climate Change. Cambridge University Press, Cambridge

Patz JA, Gibbs HK, Foley JA, Rogers JV, Smith KR (2007) Climate change and global health: quantifying a growing ethical crisis. EcoHealth 4:397–405

Pearce DW, Cline WR, Achanta AN, Fankhauser S, Pachauri RK, Tol RSJ et al (1996) The social costs of climate change: greenhouse damage and the benefits of control. In: Bruce JP, Lee H, Haites EF (eds) Climate change 1995: economic and social dimensions-contribution of working group III to the second assessment report of the Intergovernmental Panel on Climate Change. Cambridge University Press, Cambridge, pp 179–224

Rhodes CJ (2019) Sci Prog 102(1):73–89

Shaaban M, Petintin JO (2014) Renewable energy potentials in Nigeria: meeting rural energy needs. Renew Sustain Energy Rev 29:72–84

Silva RVD, Rodrigues LA, Maurício GSD, Marco AS, Bruna GD, Martins MA (2019) Biochar and mucuna increase papaya plant growth and nutrition, as well as soil fertility. Pesqui Agropecu Trop 49(e55210):1–10

Smith JL, Collins HP, Bailey VL (2010) The effect of young biochar on soil respiration. Soil Biol Biochem 42:2345–2347

Sohi S, Lopez-Capel E, Krull E, Bol R (2009) Biochar, climate change and soil: a review to guide future research. CSIRO Land and Water science report, 5/09. CSIRO Land and Water, Canberra, p 02

Sojinu OSS, Wang JZ, Sonibare OO, Zeng EY (2010) Polycyclic aromatic hydrocarbons in sediments and soils from oil exploration areas of the Niger Delta, Nigeria. J Hazard Mater 174(1–3):641–647

Sonibare J, Adebiyi F, Obanijesu E, Okelana O (2010) Air quality index pattern around petroleum production facilities. Manag Environ Qual Int J 21(3):379–392

Swanson J (2013) Climate-change mitigation potential of biochar: a review and framework for carbon accounting. MSc thesis, Duke Environmental Leadership Master of Environmental Management, Nicholas School of Environment, Duke University, Durham, pp 1–39

Tawari CC, Abowei JFN (2012) Air pollution in the Niger Delta area of Nigeria. Int J Fish Aquat Sci 1:94–117

Turkdogan MK, Fevzi K, Kazim K, Ilyas T, Ismail U (2003) Heavy metals in soil, vegetables and fruits in the endemic upper gastrointestinal cancer region of Turkey. Environ Toxicol Pharmacol 13:175–179

Ubani EC, Onyejekwe IM (2013) Environmental impact analyses of gas flaring in the Niger Delta region of Nigeria. Am J Sci Ind Res 4(2):246–252

Ujowundu CO, Nwaogu LA, Ujowundu FN, Belonwu DC (2013) Effect of gas flaring on the phytochemical and nutritional composition of *Treculia africana* and *Vigna subterranean*. Br Biotechnol J 3(3):293–304

Ukala E (2011) Gas flaring in Nigeria's Niger Delta: failed promises and reviving community voices. J Energy Clim Environ 97:97–126

UNFCCC (2015) Adoption of the Paris Agreement. https://unfccc.int/resource/docs/2015/cop21/eng/l09.pdf. Accessed 5 Nov 2019

Wang T, Camps-Arbestain M, Hedley M, Bishop P (2012) Predicting phosphorous bioavailability from high-ash biochars. Plant Soil 357:173–187

Weyant CL, Shepson PB, Subramanian R, Cambaliza MOL, Heimburger A, Mccabe D, Baum E, Stirm BH, Bond TC (2016) Black carbon emissions from associated natural gas flaring. Environ Sci Technol 500:2075–2081

Woolf D, Amonette JE, Street-Perrott FA, Lehmann J, Joseph S (2010) Sustainable biochar to mitigate global climate change. Nat Commun 1:56

Yaduma N, Kortelainen M, Wossink A (2013) Estimating mortality and economic costs of particulate air pollution in developing countries: the case of Nigeria. Environ Resour Econ 54:361–387

Zhao S, Alexandroff A (2019) Current and future struggles to eliminate coal. Energy Policy 129:511–520

Zhou Y, Cheng S, Chen D, Land J, Zhao B, Wei W (2014) A new statistical approach for establishing high-resolution emission inventory of primary gaseous air pollutants. Atmos Environ 94:392–401

# 3

# Adaptation to Climate Change: Opportunities and Challenges from Zambia

Jonty Rawlins and Felix Kanungwe Kalaba

## Contents

Introduction
- Climate Change in Southern Africa and Zambia8

Methodology .

Climate Change Risk Pathways in Zambia
- Infrastructure Risk .
- Socioeconomic Risk
- Human Health Risk
- Agricultural Risk
- Environmental Change Risk
- Cumulative Risks and Positive Feedback Loops .

Barriers to Adaptation and Resilience Building
- Poor Livelihood and Economic Diversification .
- High Rates of Land Degradation .
- Low Levels of Technical and Technological Capacities
- Inadequate Infrastructure
- Entrenched Culturally and/or Policy-Driven Practices
- Weak Policy Implementation
- Persistent Inequalities
- Cumulative Barriers and Positive Feedback Loops .

Toward Improved Climate Change Adaptation .
- Maladaptation Potential
- Is Livelihood and Economic Diversification the Answer?

---

J. Rawlins (✉)
Natural Resources, OneWorld Sustainable Investments, Cape Town, South Africa
e-mail: jrawlins@outlook.com

F. K. Kalaba
School of Natural Resources, Copperbelt University, Kitwe, Zambia

Conclusions and Recommendations
References

## Abstract

Context appropriate adaptation interventions and strategies that respond directly to localized climate change stressors, hazards, and vulnerabilities are critical for the sustainable development of countries like Zambia. This chapter examines both localized and systemic climate change risk pathways and barriers to adaptation action in Zambia.

A three-staged methodology was applied, combining content analysis, focus group discussions, and expert interviews. Livelihood diversification was identified as the central adaptation option across Zambia, despite little empirical research detailing possible risks of diversification. The dominant adaptation discourse is focused specifically on diversifying within agriculture-based livelihoods. However, as all agricultural activities are impacted by climate change, diversification also needs to be explored in value-added or alternative sectors. With this, a weak policy framework and enabling environment are exacerbating cycles of poverty that underpin climate change vulnerability in Zambia. Moreover, maladaptation risks of existing diversification interventions are high as generic approaches often do not provide suitable options to complex and localized risk profiles.

To implement a sustainable transition toward climate resilient and compatible development in Zambia, the authors recommend that a systematic livelihood diversification strategy should be rolled out and future research programs designed to support this. Specifically, this necessitates a system-wide analysis of pre-identified livelihood diversification pathways that can be adapted to different scenarios given the current and future climate uncertainties at local scales. The approach should focus on harnessing the positive feedback loops for systematic change to build resilience, while minimizing the dominant risk pathways and eliminating persistent barriers that enable positive feedback loops driving vulnerability to climate change. Thorough stakeholder engagement and incremental development of diversification options, incentives, penalties, and other governance and/or policy mechanisms will be needed to support these processes.

## Keywords

Climate change · Adaptation · Sustainable development · Diversification · Zambia

## Introduction

There is increasing evidence of the existing and potential impacts of climate change on ecosystems and society (Fischer 2019; Grüneis et al. 2018). These impacts are projected to increase in southern Africa due to projected increases in exposure and existing high levels of sensitivity and low levels of adaptive capacity (IPCC 2018). Climate change poses complex challenges to development planning and management at various scales (national, regional, and local) (Whitney and Ban 2019) through a multitude of risk pathways. Globally, enabling effective and efficient adaptation interventions has become a major policy goal over the last decade (Fischer 2019). Recent evidence suggests that adaptation strategies and policies are often developed using a top-down approach (Grüneis et al. 2018). However, there are increasingly urgent calls for locally appropriate adaptation strategies that respond to local scale climate hazards, stressors and vulnerabilities and take cognizance of localized socioeconomic and environmental conditions. Conventional public policy and sustainable development strategies do not effectively incorporate climate change impacts and projections (Whitney and Ban 2019) at a localized scale, often resulting in unintended negative consequences. Designing and implementing effective adaptation actions requires the active participation of local practitioners in order to make decisions on adaptation choices and activities that are locally relevant, credible, and contextually appropriate.

A variety of adaptation measures have been proposed across Africa. However, there is little empirical evidence of how adaptation decisions are contextually developed and to what degree local institutions participate and/or drive such processes. Local institutions understand local contexts and the various climate hazards within their regions and therefore need to be key decision-makers in the development of adaptation strategies and policies (Fischer 2019; Kupika et al. 2019). This is evidenced by extensive scholarship exploring the pervasiveness of maladaptation (e.g., Juhola et al. 2016; Magnan et al. 2016) and issues of scale and applicability of generic adaptation investments under deep uncertainty (Haasnoot et al. 2019; Hallegatte et al. 2012). This is of particular importance in developing adaptation responses to pursue sustainable development outcomes, taking into consideration ecological and social coping and adaptation capacities at multiple spatial scales.

Effective adaptation minimizes negative impacts and maximizes the potential benefits of climate change (Whitney and Ban 2019) given the resource constraints of a particular socio-ecological system. Thus, effective local strategies will, to a large extent, depend on the understanding and perceptions of local actors about opportunities to inform adaptation strategies (Whitney and Ban 2019). Thus, it is clear that pragmatic adaptation actions demand participation of all key local actors and institutions.

This chapter examines the risks from climate change that are currently being experienced and further explores current and future adaptation strategies and barriers to adaptation. Complex and cumulative climate change risk pathways and key barriers to adaptation and resilience building are explored through analyzing perceptions of local actors and institutions within the broader policy and environmental context of Zambia. The overarching objective of this research is to assess and recommend context appropriate and systemic adaptation interventions and strategies that respond directly to both national and localized climate change risks and vulnerabilities.

## Climate Change in Southern Africa and Zambia

Southern Africa is one of the global climate change hotspots (IPCC 2018). Anthropogenically driven temperature rise is at approximately twice the average global rate of temperature increase (Weber et al. 2018). This is combined with the observed increased frequency and intensity of extreme events such as droughts, heavy rainfall events, heat waves, and floods over the past few decades (Paeth et al. 2010; Taylor et al. 2017). Climate projections for the region also indicate rises in temperatures, an increase in hot days, heat waves, and heat stress as well as continued increases in the frequency and magnitude of extreme events such as droughts and floods (Engelbrecht et al. 2015; Dosio 2017; Maúre et al. 2018).

The severity of these impacts will be magnified by low levels of adaptive capacity, stemming from a plethora of socioeconomic and environmental challenges facing the Southern African region. These factors include rapid population growth, natural resource-related poverty and inequality, regional migration, significant infrastructure deficits, low levels of access to clean water and sanitation, and inefficient food, and energy systems among others (Nhamo et al. 2018; Rawlins et al. 2018).

Zambia is no exception to these trends, with severe climate change impacts projected for the economy, local livelihoods, and the broader environment (Kaluba 2014; World Bank 2019). The country is particularly vulnerable to floods and droughts, which in the past 30 years have cost the country an estimated 13.8bn USD, equivalent to around 0.4% of GDP (Kaluba 2014). Zambia has also seen decreased rainfall by an average rate of 1.9 mm/month (2.3%) per decade since 1960, and the rainfall season has become shorter and more difficult to predict – with fewer, but more intense rainfall events (AfDB 2019; ZINDC 2015). Mean annual runoff is estimated to decrease by ~13%, under future climate change scenarios (Hamududu and Ngoma 2018). This, combined with increasing climate variability, is reducing water availability across the country. Moreover, vulnerability to climate extremes such as droughts and floods is also increasing, posing a significant threat to water security in the country.

Future scenarios for Zambia over the period 2070–2099 indicate that temperatures will increase by 2 °C and rainfall could decrease by 8–10% (Bwalya 2010) with an indication of heavy rainfall events becoming more frequent in the northern highlands (AfDB 2019). As stated above, the frequency of extreme events (droughts, seasonal floods and flash floods, extreme temperatures, and dry spells), along with their intensity and magnitude, has also increased, and is already having notable knock-on effects for local livelihoods.

## Methodology

This chapter adopted a three-staged methodological approach, combining technical content analysis, focus group discussions, and semi-structured expert interviews. In the first stage, we conducted an extensive content analysis of key Zambian national policies that address adaptation to climate change and broader sustainable development in one way or another. This included the 2017 National Climate Change Policy, national sectoral policies (water, agriculture, and forestry), and national development plans (Seventh National Development plan and the Vision 2030). Analyzing these policies provided insights on policy direction at the national level and to what degree adaptation to climate change is addressed both implicitly and explicitly as well as the complex interrelations and interdependencies between adaptation and sustainable development objectives.

The second stage involved provincial workshops and a national stakeholder workshop (Lusaka, Zambia, August 2018). Provincial workshops were arranged by district sectoral leads. A total of three provincial level workshops were conducted in Western, Luapula and Northern provinces. These were useful in providing insights into key climate risks being experienced in the various districts of the provinces, and the possible adaptation interventions, which were prioritized by local participants. Participants were drawn from a variety of government departments, NGOs, private sector, and civil society organizations (CSOs). Provincial workshops had an average of 45 participants. We further conducted key informant interviews with provincial heads to understand how localized climate risks in each province related to national, provincial, and local policies and strategies for adaptation and sustainable development. These semi-structured interviews provided the necessary depth in understanding the drivers of and barriers to effective adaptation as well as how these different risks and barriers can interact and accumulate throughout socio-ecological systems. Lastly, we conducted a national workshop, which was attended by 60 experts drawn from across government departments (forestry, agriculture, environment, national development planning, and water), NGOs, and academics.

Finally, semi-structured expert interviews were carried out with experienced academics and practitioners working on a variety of climate change, natural resource, and sustainable development challenges in Zambia and Southern Africa at large. This multistaged approach has been widely used in related studies (e.g., Antwi-Agyei et al. 2018; England et al. 2018). The combination of content analysis, focus group discussions, workshops, and in-depth interviews was shown to be useful in identifying risks, barriers, and possible investments to help adapt to climate change. Specifically, the iterative development of the key findings of the analysis assisted the verification and validation of proposed systemic adaptation responses that will address both national and localized climate change risks in an integrated manner.

## Climate Change Risk Pathways in Zambia

Zambia faces a multitude of climate change risks related to not only the direct effects of climate change but also to infrastructure, social and economic conditions, human health, agriculture, and the natural environment, among others. These risks manifest through a multitude of risk pathways with a variety of impacts over different time horizons that disproportionately impact the most vulnerable populations. Additionally, cumulative impacts of these risks regularly lead to positive feedback loops that serve to magnify localized climate change risks.

## Infrastructure Risk

Heavy flooding frequently cuts road access to districts and their administrative centers. Some of the largest contributors to climate sensitivity and vulnerability in Zambia are low road density, the poor quality of many roads, the inability to traverse them for extended periods in the wet season, and the lack of maintenance and "climate proofing" of existing roads (Petrie et al. 2018).

An example of the impact of damage to infrastructure given by workshop participants was that, in remote districts, teachers may not be paid their monthly salaries for months at a time when floods cut road access. Left without an income, they must find the means to obtain food, which they may do by growing food crops themselves. This takes time that they would otherwise use for their teaching load (planning and administrative tasks), which affects the quality of education they deliver. Alternatively, teachers who find this situation difficult may not last more than 2 years in this employment and then leave, resulting in annual staff turnover as high as 30% (EPDC 2012).

Hydropower accounts for the vast majority of Zambia's energy supplies, with a number of systems currently in operation and a significant amount of planned future generation (e.g., the planned Batoka Gorge Hydropower Plant). The demand for energy is driving continued hydropower development in the region, despite the significant potential decline in hydropower output projected under a drying climate in the Zambezi Basin (Spalding-Fecher et al. 2017). The highly centralized nature and sheer size of the financial investments required to support hydropower infrastructure presents a significant fiscal risk to Zambia given these potential impacts of climate change.

Other examples of the direct and indirect impacts arising from vulnerable road infrastructure are reduced economic activity (trade cannot take place, as markets are inaccessible); and a seasonal lack of access to health and educational facilities. Finally, to compound these factors, authorities are unable to deliver aid when it is needed in times of disaster.

## Socioeconomic Risk

Flooding destroys homes as well as crops and livestock and other household assets. This has an impact on public, household, and individual health and severely affects the access of households to social services. The poor road infrastructure restricts commercial enterprise, depressing local and regional economies and increasing the socioeconomic risks of the region.

A high population growth rate is a key underlying driver of vulnerability, with the rate often exceeding the ability of the economy to absorb job seekers. This highlights the potential future risks to local economies and threatens the ability of social services such as education and health facilities to cope with the situation. High levels of education give people more options in terms of diversified livelihoods and income generating activities, while poor access to schooling and poor education or low levels of education hinders adaptation. Without education, people are forced to fall back more strongly on traditional income generation practices. Moreover, risks from the impacts of climate change are exacerbated by the lack of income diversity and high levels of poverty in general.

Gender equality is low in rural areas of Zambia and gender-based violence is common (Samuels et al. 2015). For example, early marriages lead to increased vulnerability of the affected girls. Education levels are low especially for girls who drop out of school to get married. With limited skills, women are restricted in terms of both the value they can add to available natural resources, and the ability to utilize various economic opportunities that would increase the efficiency and productiveness of households to improve their adaptive capacities.

## Human Health Risk

An unhealthy population (or one with a high burden of disease) has a lower resilience to climate change impacts that challenge human health (Smith et al. 2014). HIV/Aids, malaria, and tuberculosis are the most prevalent and therefore serious diseases in Zambia. In addition, the incidence of these diseases reduces the resilience of those individuals and communities affected by other health challenges (i.e., comorbidities).

Poor health infrastructure – inadequate resources, including too few health facilities and staff – is a significant source of vulnerability. This is exacerbated by the remoteness of many districts, and poor roads, meaning that some areas are periodically isolated from health facilities during the wet season, with its concomitant impacts of flooding. It is unclear at this stage how emerging global health risks such as the Corona Virus 2019 (COVID-19) will be influenced by climate change risks; however, it is clear that the potential knock-on effects of a national or global pandemic will further limit the available health infrastructure capacity.

Social issues that result from climate-induced stresses include impacts on mental health. Climate change can affect people directly through trauma, which is evident in climate hazards such as floods. It also affects mental health indirectly, through impacts on physical health and broader community well-being (Berry et al. 2010).

Heat waves and persistent high temperatures will not only alter disease vector habitat distribution (such as malaria and tsetse fly, see Caminade et al. (2014) and Terblanche et al. (2008)), but also pose direct threats to human health in the form of heat stroke, exhaustion, and death.

## Agricultural Risk

Flooding is hazardous to crops across Zambia, though this risk profile varies from region to region. For example, in the higher, steeper areas of districts in Muchinga

Province, flooding is less of a risk because smaller areas of crops are exposed to the hazard. However, crops along the major rivers can be exposed to prolonged periods of heavy flooding, for example, those in Western Province or parts of Luapula and Northern provinces. With a substantial portion of their food resources for a season destroyed, people must seek alternative forms of income, including relocating for at least part of the agricultural year to other regions in order to generate an income, often by selling labor, charcoal manufacture, or fishing.

The impact of drought on crops and livelihoods varies between the provinces and their districts. In general, mean annual rainfalls are lower in the western and southern regions of Zambia, and droughts are more intense and longer than elsewhere. Droughts in this region have very significant impacts on crop yield. In the northern provinces, where mean annual rainfalls are significantly higher, drought is not a significant risk.

Temperature increases above certain thresholds substantially reduce the grain yield of maize, especially in soils which have a poor water holding capacity (see, e.g., Hatfield and Prueger 2015). This effect is more likely in the western and southern provinces than in the northern provinces, which are at higher altitudes, with higher rainfall and longer growing seasons.

Storm damage to crops – a result of high windspeed during intense events, or hail, is known as a relatively localized event. While devastating for the affected householders, these usually do not occur at district or provincial level. Climate change has shown to have negative effects on livestock leading to reduction in available fodder for animals due to droughts. Floods also affect livestock by reducing the grazing land available, further increasing pressure. Disease incidence for livestock also increased with corresponding increase in animal mortality.

## Environmental Change Risk

Zambia has a burgeoning challenge in deforestation, which has been recognized for some time, in essence since the beginning of the last century (Holden 2001). There are three main causes – land clearing for "slash-and-burn" or *Chitimene* agriculture, charcoal making, and logging for high-value timber. The earlier land-use changes resulted from population growth and the *Chitimene* and *Fundikila* ("grass-mound") farming systems, which required migration and opening up of new areas, especially when cassava and maize-fertilizer systems were implemented in Northern Zambia. These entrenched practices have caused significant environmental changes across the Zambian landscape as well as increased dependence on unsustainable supplies of natural resources, both of which increase vulnerability to climate change.

Fisheries are a critical natural resource in Zambia, but are experiencing overexploitation and are under increasing pressure from external development activities that compromise river ecosystem services and functioning (Cowx et al. 2018). Although there is limited evidence directly linking the decline in inland fisheries in recent years to climate change, there are strong linkages to these declines with overfishing driven by human behavior change (Marshall 2012).

Deforestation and its relationship to climate sensitivity hinges on how changes in vegetation cover affect the physical and chemical components of the environment in ways that are likely to reduce resilience to change. Deforestation, followed by agricultural disturbance, results in loss of soil organic carbon and of soil fertility (Chidumayo and Kwibisa 2003). This effect is enhanced by grassland fires (in areas where grass has replaced lost tree cover), with repeated burns also reducing soil carbon. Soil carbon is also important for enhancing soil water retention, an obvious necessity for plant growth and for maintaining growth potential. Deforestation also leads to a decline in soil moisture, which then leads to more surface energy being converted into heat at the land surface. This results in a hotter environment, which at these latitudes is damaging to crop growth and alters pest and disease organism life cycles. Ultimately, deforestation often leads to lower climate resilience.

### Cumulative Risks and Positive Feedback Loops

Infrastructure, socioeconomic, human health, agricultural, and environmental change risks of climate change are not independent of one another and more often than not serve to magnify each other (Mora et al. 2018). These positive feedback loops result in cumulative risk profiles across Zambia, that are deeply entrenched through the existing socio-ecological and economic systems.

Risks of climate change to infrastructure such as road closures and limited access to critical infrastructure serve to exacerbate human health risks because people cannot access hospitals and clinics. Poor health outcomes have knock-on impacts for socio-economic livelihood activities by reducing the capacity of individuals and households to generate income and/or support social activities within a household. Such risks are closely tied to agriculture as this is the dominant livelihood activity in Zambia; for example, if people cannot work due to poor health or are forced to travel far distances to seek medical care they are often not able to tend to their crops or livestock. This further magnifies the potential risks from climate change to agricultural activities and ultimately agricultural output. Deforestation and over-harvesting of valuable and limited natural resources such as inland fisheries often result from crop failure or livestock die offs, further increasing the risks related to environmental change.

## Barriers to Adaptation and Resilience Building

There are a number of environmental, social, economic, and political barriers to effective climate change adaptation in Zambia. These barriers interact throughout complex socio-ecological systems to reduce adaptive capacities and compound the limits to adaptation throughout the country.

### Poor Livelihood and Economic Diversification

Most rural livelihoods are dependent on natural resources and agriculture. In other words, Zambia has not seen the structural changes in its economy that would enable a permanent shift to one dependent on secondary or tertiary economic activities such as manufacturing and services, thus transforming from a natural resource-based economy. This is strongly aligned with broader development imperatives to shift employment from low- to high-productivity sectors.

Local markets and value chains for locally produced goods (especially agricultural products) are underdeveloped. Access to markets is low, primarily due to poor road infrastructure, impacted further by floods and extreme heat, which impact transport time and/or the modality of agricultural produce. Beneficiation processes related to output from key livelihood activities for economic development are highly underdeveloped. Thus, having livelihoods focused on a few highly vulnerable sectors and natural resources serves only to concentrate risks of climate change while simultaneously limiting the ability to adapt. Ultimately, this has resulted in low levels of livelihood and economic diversification both across primary, secondary, and tertiary economic activities as well as within primary production activities themselves.

## High Rates of Land Degradation

Land degradation is primarily driven by unsustainable livelihood practices that are rooted in high levels of poverty and inequality. Harvesting trees for charcoal production and for trade (e.g., for export to China) is adding to already high levels of deforestation. Zambians are often not even the main beneficiaries of revenues from tree-based trade. *Mukula* (Rosewood), is heavily harvested for timber exports to China. Local cutters receive an average price of about US$23/cubic meter of timber harvested, while manufacturers in China pay around US$1,000/cubic meter to importers, before any further processing is done on the exported logs (Cerutti et al. 2018).

Populations are likely to continue to diversify their livelihoods through these and other activities, such as illegal poaching, until there is a proper incentive to secure their livelihoods more sustainably. Together with climate change impacts, these activities are resulting in widespread land degradation and loss of ecosystems, biodiversity, and services that are critical to human welfare and livelihoods, such as water, fisheries, and forests.

## Low Levels of Technical and Technological Capacities

This aspect is evident in very low levels of adoption of new technologies, which range from alternative farming practices, such as conservation agriculture, to capacities to build new climate resilient infrastructure or to retrofit existing infrastructure to be climate resilient. The shift toward decentralization of governance in Zambia raises the challenge of meeting technical capacity needs across the levels of implementation. With this, districts across the studied provinces are limited by lack of the technical capacities needed to drive this implementation. There is thus an evolving trade-off between the necessity to decentralize decision-making power and governance structures with meeting the technical needs for livelihood and economic diversification at the local level to adapt to existing and future impacts of climate change.

## Inadequate Infrastructure

Zambia's high levels of infrastructure deficits in transport, energy, Information and Communications Technology (ICT), and water and sanitation significantly impede adaptive capacities. Notably, business productivity is severely hampered by the infrastructure deficits. This reduces investments in commercial activities and limits expansion of smaller businesses, further limiting income diversification opportunities. Furthermore, 50% of the productivity gap in Zambia is attributed by firms to the state of the country's infrastructure (Foster and Dominguez 2011). Most of the productivity gap is attributed to power, followed by transport and ICT. As mentioned above, road infrastructure is insufficient and plays a pivotal role for all forms of infrastructure.

## Entrenched Culturally and/or Policy-Driven Practices

Zambians are often forced to default to livelihoods and coping strategies that are supported by long-standing traditional beliefs, which are in some instances supported by well-entrenched policies. For example, high cultural value is placed on cattle ownership, to the extent that farmers often refuse to sell them off during extreme droughts, in the hopes they will survive and regain health, even though the farmers cannot feed and water them. Often the cattle do not survive.

Policies can also entrench practices not conducive to climate change. For example, policy-driven maize production has increased farmer dependence on a vulnerable livelihood and its status as the country's most desired food staple (Chapoto et al. 2010). Problematically, maize is not a drought-resilient crop and yields in Zambia are quite low, also partly due to the policy emphasis on production rather than productivity (Petrie et al. 2014). Consequently, farmers are not diversifying their risks by growing additional or alternate crops, which may be more drought resilient. Nor are they considering alternate livelihoods because the culturally and policy-driven practice of maize farming is so entrenched.

## Weak Policy Implementation

Supportive policy frameworks that enable adaptation are relatively well established in Zambia. For example, the 2017 National Climate Change Policy details a broad approach to strengthen the implementation of adaptation measures to reduce vulnerability to climate change, which supports and builds on the objectives outlined in the 7th National Development Plan. However, these policies are generally not effectively applied or fully implemented due to a number of related barriers, such as breadth of technical capacity, conflicting discourse on development objectives, lack of financing opportunities, and weak institutional arrangements (Romdhani et al. 2018). This is particularly true in rural areas, where institutional arrangements and related enforcement mechanisms are relatively weak and government institutions are poorly capacitated; yet people in these areas are the most vulnerable to the impacts of climate change.

## Persistent Inequalities

Gender inequality and gender-based violence persists despite transversal policy. Women lack access to land in comparison to men, under customary law that disadvantages women in terms of land allocation. This compromises agricultural productivity and food security, particularly in situations where women have the responsibility over daily household welfare without decision-making powers over land and land use, and in the face of little access to disposable income. With this, female farmers tend to have even lower access than men to agricultural inputs.

Even if there were greater opportunities for livelihoods and incomes beyond natural resource-based sectors, education, a key indicator of adaptive capacity, is low in the studied provinces. This limits people's access to opportunities, particularly affecting women. Inequalities are also highly evident in access to safe water, electricity, and public health services, which tends to be much lower in rural areas making people less able to adapt to the impacts of climate change. Ultimately, this reduces the adaptive capacity of these individuals and communities and reduces their opportunities to diversify.

## Cumulative Barriers and Positive Feedback Loops

Similar to the cumulative nature of the climate change risk pathways, the abovementioned barriers to adaptation can also create positive feedback loops that serve to reinforce these barriers and ultimately limit resilience building opportunities (Shackleton et al. 2015). A key example from the Zambian context relates to both weak policy implementation and the myopic culturally entrenched and policy-driven practices. Agriculture-based livelihoods being culturally entrenched and together driven by centralized blanket maize cultivation incentives create more significant barriers to effective adaptation when combined with poor implementation of climate change policies. Moreover, traditional livelihoods serve to reinforce the status quo around traditional gender roles and thus further entrench the persistent inequalities. Moreover, traditional livelihood practices and high levels of inequality and poverty drive unsustainable land-use practices and ultimately lead to degradation. This phenomenon is heightened due to the increasing nature of crop failures and livestock die offs due to the impacts of climate change.

Given that low population densities and high travel distances necessitate greater per capita investment for infrastructure construction (Chapman et al. 2013), the widespread nature of rural and subsistence-based livelihoods across the country compound the difficulty of infrastructure provision. The same logic applies to the adoption of novel technologies and livelihood practices as well as the opportunities to access alternative markets and economies.

## Toward Improved Climate Change Adaptation

Climate change risk pathways and barriers to adaptation manifest through a complex network of social, economic, and environmental feedback loops, often resulting in positive feedback loops that intensify the initial perceived risk of a particular climate hazard or stressor. These potentially antagonistic relationships between different risk pathways and socioeconomic barriers to adaptation highlight the complexity of the interdependent nature of socio-ecological systems in Zambia. The benefits of developing adaptation responses that focus on individual climate change risk pathways or barriers will be abated due to the complex synergistic and antagonistic feedback loops that characterize these systems. Thus, these complex interactions highlight the need for systematic responses to minimize risks and overcome the barriers.

In Zambia, the complex and emergent properties of the socio-ecological systems, combined with their strong dependency on climate systems, elucidate the intrinsic and inextricable relationship between climate change and sustainable development. Thus, effective adaptation responses need to take a systematic approach that attempts to address the myriad of complex climate change risks concurrently with existing barriers to adaptation. Kim et al. (2018) show that community social and ecological climate change risk characteristics were reduced by higher levels of social capital combined with local proactive planning and policy measures focused on systematic responses to cumulative exposure. The notion of an integrated and systematic response to stressors and shocks on socio-ecological systems and related economic activities is widely considered effective, whether or not these external stressors and shocks are related to climate change or not (Cinner and Barnes 2019; Fedele et al. 2019; Hafezi et al. 2018).

### Maladaptation Potential

Responding to a myriad of climate change threats across Zambia requires both increased resilience to short- to medium-term shocks (e.g., floods and droughts) combined with sustained and more gradual adaptation to climate change (temperature rise and seasonal rainfall changes) across systems and society. Moreover, this necessitates rigorous assessments of maladaptation potential for proposed interventions. Much of the onus for this falls on government institutions and the Zambian governance framework because the role of the developmental state is to balance social development with economic growth in a sustainable manner. However, maladaptation risks are high and generic approaches to diversifying livelihoods do not always provide the best options to complex risk profiles associated with poverty, climate change, and environmental degradation.

Maladaptive actions may lead to increased risks of adverse climate-related outcomes (IPCC 2018). Thus, understanding the potential for maladaptation is critical for planning and budgeting for adaptation interventions. Zambia faces enormous opportunities to improve livelihoods and living standards through adaptation; however, the significant scope of the opportunity raises the likelihood of maladaptive outcomes.

Improving road infrastructure, for example, has its own risks, despite the notable potential for positive sustainable development and climate change adaptation outcomes. In many parts of Zambia, such as Western Province, illegal tree harvesting of species such as Mzauli (African rosewood – *Guibourtia coleosperma*) is taking place, with the timber being exported through Namibia. This is leading to a decline in this species. Improved roads into these areas could facilitate increased penetration of illegal tree fellers.

## Is Livelihood and Economic Diversification the Answer?

Livelihood and economic diversification are regularly identified as a key adaptation option given the current socioeconomic state of Zambian populations and communities and the current and projected impacts of climate change (Petrie et al. 2018). The dominant discourse around adaptation in Zambia is focused on agriculture-based livelihoods, which are facing the greatest risks from climate change. Currently, in response, adaptation interventions are focused on transitions to (and between) different forms of conservation agriculture, crop diversification, and a combination of farm and nonfarm activities such as seasonal fishing and charcoal production (Loison 2015; Romdhani et al. 2018).

Zambia's favorable agro-climatic conditions are suitable for producing a diverse variety of livestock and crops; however, legacy blanket policies promoting the cultivation and consumption of maize have limited these opportunities significantly. In 2015, ±98% of smallholder households were cultivating maize, while it is estimated that more than half of the land under crop production in Zambia is devoted to maize (CSO 2015). Moreover, Mofya-Mukuka and Hichaambwa (2018) found that government input subsidies and direct market support for maize have negatively affected crop diversification by rural households. Additionally, compared to neighboring countries in the region, Zambia's level of crop diversification is low (as measured by the Simpson Index of Diversification = ±0.37) (World Bank 2018). Despite this the potential for diversification has been shown to be good. Alfani et al. (2019) found in Zambia that general diversification actions such as livestock and income diversification and the adoption of agro-forestry as an alternative economic activity served to moderate the impact of a recent severe drought. Interestingly, it was also found that some agro-ecological practices such as minimum soil disturbance was not effective in moderating any yield and income effects of the drought. Additionally, Wineman and Crawford (2017) demonstrate that given the current set of available diversification options, autonomous on-farm adaptation will not be sufficient to offset the negative yield effects of climate change.

Combined with the notable vulnerability of maize production to projected climate change in Zambia (Wineman and Crawford 2017; Petrie et al. 2014), the rationale for diversification to improve income and/or food security is clear. However, the perceived benefits of diversification might not always materialize on the ground, as the enabling environment to support such diversification is weak. Loison (2015) emphasizes that it is only the relatively better-off smallholders with sufficient assets who achieve successful livelihood diversification; and thus, associated increases in income and wealth based on livelihood diversification are yet to show notable benefits for smallholder farmers that are implementing diversified livelihood activities. The same argument holds in relation to improving climate resilience through livelihood diversification. Climate change vulnerability in Zambia is largely a function of people's poor ability to adapt (i.e., adaptive capacity), which is driven by high levels of poverty and a lack of access to basic services and supporting infrastructure (Petrie et al. 2018).

Despite multiple proponents of diversification as a key strategy for climate-smart agriculture in Zambia (Arslan et al. 2018), there is little empirical research detailing the possible risks of diversification. Moreover, risks of promoting diversification are exacerbated by the weak enabling environment. For example, diversifying agricultural practices requires knowledge, skills, and inputs to grow multiple types of crops and/or rear multiple different types of livestock. Not only could this cost more in the short run; there is also a greater risk of making mistakes. Diversification can also lead to localized competition for markets, leading to possible conflicts. Importantly, promoting over-diversified livelihood activities might limit opportunities to upscale by taking advantage of economies of scale through access more well developed markets and value chains. These risks might exacerbate the cycles of poverty that underpin climate change vulnerability in Zambia.

It is clear that livelihood diversification presents significant opportunities and challenges, which means it should not be viewed as an adaptation panacea in low income settings. People will naturally adapt and diversify, where livelihood and/or income opportunities are better. Hence there is a clear need for dedicated policy and governance efforts to systematically target, support, and incentivize context-specific livelihood diversification opportunities that are harmonized across competing developmental objectives. Bhatta et al. (2015) highlight that livelihood and diversification strategies need to be tailored to both localized climatic and non-climatic resources alike to minimize risks and facilitate the lowest-cost (social and economic) transition to climate resilience.

## Conclusions and Recommendations

The system- and society-wide impacts of climate change in Zambia require transformational change in sustainable development. Zambia's dependence, for example, on agriculture for its economy and for livelihoods, places the country and its populations at high risk to climate change. The related national policies, such as the incentivization of maize production, do not capture the localized livelihood contexts and climate risk profiles. The link between climate change and sustainable development in Zambia is inextricable, however, and thus these two concepts should not be viewed in isolation from one another.

Livelihood diversification was identified as the central adaptation option across Zambia, despite little empirical research detailing possible risks of diversification. The dominant adaptation discourse is focused specifically on diversifying within agriculture-based livelihoods. However, as all agricultural activities are impacted by climate change, diversification also needs to be explored in value added or alternative sectors. With this, a weak policy framework and enabling environment is exacerbating cycles of poverty that underpin climate change vulnerability in Zambia. Moreover, maladaptation risks of existing diversification interventions are high as generic approaches often do not provide suitable options to complex and localized risk profiles.

It is evident that there is a strong need to diversify both within and outside of the agriculture sector, but how this is done will be critical. Thus, the authors recommend the development of a detailed spatial assessment of the economic, environmental, and social complexities that contribute to differing levels of climate change vulnerability throughout Zambia and how these contribute to or detract from realizing national sustainable development objectives. Specifically, there is a need to focus on harnessing the positive feedback loops for systematic change to build resilience, while minimizing the dominant risk pathways and eliminating persistent barriers that enable positive feedback loops driving vulnerability to climate change.

First, this will necessitate a detailed examination of a selected range of pathways for alternate crops, livestock farming, fisheries exploitation, and secondary and tertiary services to understand key elements ranging from production to market economics across the various value chains. In this way it will be possible to establish which livelihoods are most appropriate for different areas, and how these might be influenced by projected changes in climate. It will also be important to ascertain how these can best be initiated and sustained from initial development onward in a way that aligns with national development objectives and policy priorities. Potential maladaptation risks of diversification can be addressed through continued stakeholder engagement and incremental development of diversification incentives, penalties, and other governance and/or policy support mechanisms.

To ensure a sustainable transition toward climate resilient and compatible development in Zambia, this systematic approach toward facilitating diversification throughout the economy should be supported by complementary approaches toward land-use and natural resources management, urban planning and infrastructure development. Future research endeavors in the region should focus on supporting this type of holistic approach by developing the evidence-base for change. Specifically, there is a need for a greater understanding of the complex interactions and emergent properties of socio-ecological systems in low-income settings that can inform policy direction at the local scale. This applies specifically to the derivation and classification of combined climate change and developmental risks, how these accumulate in a system and how best these can be addressed given existing resource constraints.

Additionally, future research should also focus on developing innovative densified development models in low-income settings, which can balance the need to simultaneously realize social, economic, and environmental development imperatives. Particular emphasis should be given to the potential role of nature-based solutions and the integration of ecological and traditional gray infrastructure. Ultimately, this approach toward evidence-led policy implementation will assist to minimize risks of upscaling livelihood and economic diversification actions, while continuing to facilitate an effective transition toward climate resilience through planned and incremental change.

## References

AfDB (2019) National climate change profile: zambia. Available: https://www.afdb.org/en/documents/zambia-national-climate-change-profile

Alfani F, Arslan A, McCarthy N, Cavatassi R, Sitko N (2019) Climate-change vulnerability in rural Zambia: the impact of an El Niño-induced shock on income and productivity. FAO agricultural development economics working paper 19-02. FAO, Rome, p 41

Antwi-Agyei P, Dougill A, Agyekum T, Stringer L (2018) Alignment between nationally determined contributions and the sustainable development goals for West Africa. Clim Pol 18:1–17. https://doi.org/10.1080/14693062.2018.1431199

Arslan A, Cavatassi R, Alfani F, Mccarthy N, Lipper L, Kokwe M (2018) Diversification under climate variability as part of a CSA strategy in rural Zambia. J Dev Stud 54:457–480. https://doi.org/10.1080/00220388.2017.1293813

Berry HL, Bowen K, Kjellstrom T (2010) Climate change and mental health: a causal pathways framework. Int J Public Health 55(2):123–132

Bhatta GD, Aggarwal PK, Shrivastava A (2015) Livelihood diversification and climate change adaptation in Indo-Gangetic plains: implication of rainfall regimes. J Agric Environ 16:77–94. https://doi.org/10.3126/aej.v16i0.19841

Bwalya SM (2010) Climate change in Zambia: opportunities for adaptation and mitigation through Africa bio-carbon initiative. Center for International Forest Research Southern Africa, Lusaka

Caminade C, Kovats S, Rocklov J, Tompkins AM, Morse AP, Colón-González FJ, Stenlund H, Martens P, Lloyd SJ (2014) Impact of climate change on global malaria distribution. PNAS 111:3286–3291. https://doi.org/10.1073/pnas.1302089111

Cerutti PO, Gumbo D, Moombe K, Schoneveld G, Nasi R, Bourland N, Weng X (2018) Informality, global capital, rural development and the environment: Mukula (rosewood) trade between China and Zambia. Research report. IIED/CIFOR, London/Lusaka

Chapman L, Azevedo JA, Prieto-Lopez T (2013) Urban heat & critical infrastructure networks: a viewpoint. Urban Clim 3:7–12. https://doi.org/10.1016/j.uclim.2013.04.001

Chapoto A, Govereh J, Haggblade S, Jayne TS (2010) Staple food prices in Zambia. Food Security Collaborative working papers 58556. Michigan State University, Department of Agricultural, Food, and Resource Economics

Chidumayo EN, Kwibisa L (2003) Effects of deforestation on grass biomass and soil nutrient status in miombo woodland, Zambia. Agric Ecosyst Environ 96:97–105. https://doi.org/10.1016/S0167-8809(02)00229-3

Cinner JE, Barnes ML (2019) Social dimensions of resilience in social-ecological systems. One Earth 1:51–56. https://doi.org/10.1016/j.oneear.2019.08.003

Cowx IG, Lungu A, Kalonga M (2018) Optimising hydropower development and ecosystem services in the Kafue River, Zambia. Mar Freshw Res 69:1974–1982. https://doi.org/10.1071/MF18132

CSO (2015) Rural Agricultural Livelihood Survey (RALS). Central Statistical Office, Ministry of Agriculture and Cooperatives and Indaba Agricultural Policy Research Institute. CSO, MAL, IAPRI, Lusaka

Dosio A (2017) Projection of temperature and heat waves for Africa with an ensemble of CORDEX Regional Climate Models. Clim Dyn 49(1):493–519. https://doi.org/10.1007/s00382-016-3355-5

Engelbrecht F, Adegoke J, Bopape MJ, Naidoo M, Garland R, Thatcher M, McGregor J, Katzfey J, Werner M, Ichoku C, Gatebe C (2015) Projections of rapidly rising surface temperatures over Africa under low mitigation. Environ Res Lett 10:085004. https://doi.org/10.1088/1748-9326/10/8/085004

England MI, Dougill AJ, Stringer LC, Vincent KE, Pardoe J, Kalaba FK, Mkwambisi DD, Namaganda E, Afionis S (2018) Climate change adaptation and cross-sectoral policy coherence in southern Africa. Reg Environ Change 18:2059–2071. https://doi.org/10.1007/s10113-018-1283-0

EPDC (2012) Zambia – making sense of data to improve education for development. Education Policy Data Center. https://www.epdc.org/country/zambia

Fedele G, Donatti CI, Harvey CA, Hannah L, Hole DG (2019) Transformative adaptation to climate change for sustainable social-ecological systems. Environ Sci Pol 101:116–125. https://doi.org/10.1016/j.envsci.2019.07.001

Fischer AP (2019) Adapting and coping with climate change in temperate forests. Glob Environ Chang. https://doi.org/10.1016/j.gloenvcha.2018.10.011

Foster V, Dominguez C (2011) Zambia's infrastructure: a continental perspective. The World Bank policy research working paper 5599. The World Bank, Washington, DC

Grüneis H, Penker M, Höferl KM, Schermer M, Scherhaufer P (2018) Why do we not pick the low-hanging fruit? Governing adaptation to climate change and resilience in Tyrolean mountain agriculture. Land Use Policy 79:386–396

Haasnoot M, van Aalst M, Rozenberg J, Dominique K, Matthews J, Bouwer LM, Kind J, Poff NL (2019) Investments under non-stationarity: economic evaluation of adaptation pathways. Clim Chang 161:1–13

Hafezi M, Sahin O, Stewart RA, Mackey B (2018) Creating a novel multi-layered integrative climate change adaptation planning approach using a systematic literature review. Sustainability 10:4100. https://doi.org/10.3390/su10114100

Hallegatte S, Shah A, Lempert R, Brown C, Gill S (2012) Investment decision making under deep uncertainty-application to climate change. The World Bank, Washington, DC

Hamududu B, Ngoma H (2018) Impacts of climate change on water availability in Zambia: implications for irrigation development. Technical paper number 7. Indaba Agricultural Policy Research Institute (IAPRI), Lusaka

Hatfield JL, Prueger JH (2015) Temperature extremes: effect on plant growth and development. USDA Research and Programs Extreme Events. Weather Clim Extremes 10:4–10. https://doi.org/10.1016/j.wace.2015.08.001

Holden S (2001) A century of technological change and deforestation in the miombo woodlands of Northern Zambia. In: Angelsen A, Kaimowitz D (eds) Agricultural technologies and tropical deforestation. CABI, New York, pp 251–270

IPCC (2018) Global warming of 1.5 °C: an IPCC special report on the impacts of global warming of 1.5 °C above pre-industrial levels and related global greenhouse gas emission pathways, in the context of strengthening the global response to the threat of climate change, sustainable development, and efforts to eradicate poverty. Intergovernmental Panel on Climate Change, Geneva

Juhola S, Glaas E, Linnér BO, Neset TS (2016) Redefining maladaptation. Environ Sci Pol 55:135–140

Kaluba D (2014) Republic of Zambia pilot program for climate resilience: climate resilience and economic development. http://www.oecd.org/env/cc/Session%204%20-%20Kaluba%20Climate%20Resilience%20and%20Economic%20Growth.pdf

Kim H, Marcouiller DW, Woosnam KM (2018) Rescaling social dynamics in climate change: the implications of cumulative exposure, climate justice, and community resilience. Geoforum 96:129–140. https://doi.org/10.1016/j.geoforum.2018.08.006

Kupika LO et al (2019) Local ecological knowledge on climate change and ecosystem-based adaptation strategies promote resilience in the Middle Zambezi Biosphere Reserve, Zimbabwe. Wildlife conservation in tropical savanna ecosystem. Scientifica 2019:3069254. https://doi.org/10.1155/2019/3069254

Loison SA (2015) Rural livelihood diversification in sub-Saharan Africa: a literature review. J Dev Stud 51:1125–1138. https://doi.org/10.1080/00220388.2015.1046445

Magnan AK, Schipper E, Burkett M, Bharwani S, Burton I, Eriksen S, Gemenne F, Schaar J, Ziervogel G (2016) Addressing the risk of maladaptation to climate change. WIREs Clim Change 7:646–665. https://doi.org/10.1002/wcc.409

Marshall BE (2012) Does climate change really explain changes in the fisheries productivity of Lake Kariba (Zambia-Zimbabwe)? Trans R Soc S Afr 67:45–51. https://doi.org/10.1080/0035919X.2012.694083

Maúre G et al (2018) The southern African climate under 1.5 °C and 2 °C of global warming as simulated by CORDEX regional climate models. Environ Res Lett 13(6):065002. https://doi.org/10.1088/1748-9326/aab190

Mofya-Mukuka R, Hichaambwa M (2018) Livelihood effects of crop diversification: a panel data analysis of rural farm households in Zambia. Food Sec 10:1449–1462

Mora C, Spirandelli D, Franklin EC, Lynham J, Kantar MB, Miles W, Smith CZ, Freel K, Moy J, Louis LV, Barba EW, Bettinger K, Frazier AG, Colburn Ix JF, Hanasaki N, Hawkins E, Hirabayashi Y, Knorr W, Little CM, Emanuel K, Sheffield J, Patz JA, Hunter CL (2018) Broad threat to humanity from cumulative climate hazards intensified by greenhouse gas emissions. Nat Clim Chang 8:1062–1071. https://doi.org/10.1038/s41558-018-0315-6

Nhamo L, Ndlela B, Nhemachena C, Mabhaudhi T, Mpandeli S, Matchaya G (2018) The water-energy-food nexus: climate risks and opportunities in southern Africa. Water 10(5):567. https://doi.org/10.3390/w10050567

Paeth H et al (2010) Meteorological characteristics and potential causes of the 2007 flood in sub-Saharan Africa. Int J Climatol 31(13):1908–1926. https://doi.org/10.1002/joc.2199

Petrie B, Chapman A, Midgley A, Parker R (2014) Risk, vulnerability and resilience in the Limpopo River Basin system: climate change, water and biodiversity – a synthesis. For the RESILIM Program, USAID. OneWorld Sustainable Investments, Cape Town

Petrie B, Rawlins J, Perkins D, Lumbroso D, Chapman A (2018) Climate change risk and vulnerability assessment in Luapula, Muchinga, Northern and Western Provinces of Zambia. OneWorld Sustainable Investments, Cape Town

Rawlins J, Petrie B, Gerhard M, Filipova A (2018) Towards cooperative transboundary environmental management in SADC: politics and harmonisation. Political drivers for sustainable ecosystems: building peace. World Water Week, August 2018, Stockholm

Romdhani A, Coll Besa M, Scott D, Bharwani S (2018) Zambia's adaptation to climate change. Stockholm Environment Institute, Stockholm

Samuels F, Ndubani P, Walker D, Simbaya J (2015) Baseline study: Stamping Out and Preventing Gender Based Violence (STOP GBV) in Zambia. ODI Frontiers Research Group, ODI, Blackfriars

Shackleton S, Ziervogel G, Sallu S, Gil T, Tschakert P (2015) Why is socially-just climate change adaptation in sub-Saharan Africa so challenging? A review of barriers identified from empirical cases. WIREs Clim Change 6:321–344. https://doi.org/10.1002/wcc.335

Smith KR, Woodward A, Campbell-Lendrum D, Chadee DD, Honda Y, Liu Q, Olwoch JM, Revich B, Sauerborn R (2014) Human health: impacts, adaptation, and co-benefits. In: Field CB, Barros VR, Dokken DJ, Mach KJ, Mastrandrea MD, Bilir TE, Chatterjee M, Ebi KL, Estrada YO,

Genova RC, Girma B, Kissel ES, Levy AN, MacCracken S, Mastrandrea PR, White LL (eds) Climate change 2014: impacts, adaptation, and vulnerability. Part A: global and sectoral aspects. Contribution of working group II to the fifth assessment report of the Intergovernmental Panel on Climate Change. Cambridge University Press, Cambridge, UK/New York, pp 709–754

Spalding-Fecher R, Joyce B, Winkler H (2017) Climate change and hydropower in the Southern African Power Pool and Zambezi River Basin: system-wide impacts and policy implications. Energy Policy 103:84–97. https://doi.org/10.1016/j.enpol.2016.12.009

Taylor CM et al (2017) Frequency of extreme Sahelian storms tripled since 1982 in satellite observations. Nature 544(7651):475–478. https://doi.org/10.1038/nature22069

Terblanche JS, Clusella-Trullas S, Deere JA, Chown SL (2008) Thermal tolerance in a south-east African population of the tsetse fly *Glossina pallidipes* (Diptera, Glossinidae): implications for forecasting climate change impacts. J Insect Physiol 54:114–127. https://doi.org/10.1016/j.jinsphys.2007.08.007

Weber T et al (2018) Analysing regional climate change in Africa in a 1.5 °C, 2 °C and 3 °C global warming world. Earth's Future 6:1–13. https://doi.org/10.1002/2017ef000714

Whitney CK, Ban NC (2019) Barriers and opportunities for social-ecological adaptation to climate change in coastal British Columbia. Ocean Coast Manag 179:104808

Wineman A, Crawford EW (2017) Climate change and crop choice in Zambia: a mathematical programming approach. NJAS Wagening J Life Sci 81:19–31. https://doi.org/10.1016/j.njas.2017.02.002

World Bank (2018) Productive diversification of African agriculture and its effects on resilience and nutrition. World Bank, Washington, DC

World Bank (2019) Climate change knowledge portal: zambia. Available: https://climateknowledgeportal.worldbank.org/country/zambia

ZINDC (2015) Zambia's Intended Nationally Determined Contribution (INDC) to the 2015 Agreement on Climate Change. Available: http://www4.unfccc.int

4

# Climate Change Adaptation: Opportunities for Increased Material Recycling Facilities in African Cities

Gamuchirai Mutezo, Jean Mulopo and Dumisani Chirambo

## Contents


Introduction
- Towards Zero Carbon Cities
- Objectives and Chapter Outline

Literature Review
- Evolution of Sustainable Urban Planning in African Cities
- Climate Change Adaptation
- State of Waste Management in Africa
- Material Recycling Facilities

Research Approach

Case Studies
- Urbanization Dynamics
- Urban Waste Impediments
- Current and Proposed Strategies
- Role of Informal Waste Collectors
- Nationally Determined Contributions

Proposed Recommendations and Conclusion

Reference



G. Mutezo (✉) · J. Mulopo
School of Chemical and Metallurgical Engineering, Faculty of Engineering and Built Environment, University of the Witwatersrand, Johannesburg, South Africa
e-mail: mutezo.gamuchirai@gmail.com; Jean.Mulopo@wits.ac.za

D. Chirambo
Seeds of Opportunity, Blantyre, Malawi
e-mail: sofopportunity@gmail.com; info@seedsofopportunity.org

**Abstract**

Africa's urban morphology is expected to develop at a steady rate between 2020 and 2050. Population growth, rising urbanization rates, growing energy consumption, and industrialization are only a few of the reasons causing these changes. Likewise, waste production is projected to rise from 125 million tons in 2012 to 244 million tons annually by 2025. Around 60.0% and 80.0% of African waste is made up of organic material, which is a viable methane source. Fly tipping, free disposal, landfilling, and incineration have been used as a large-scale waste treatment system in most African cities. However, with the anticipated morphological changes, these solutions are no longer viable in the future due to lack of airspace, availability of urban land for new landfill sites, and concerns over carbon emissions. This chapter discusses the potential for improved adoption of material recycling facilities (MRF) in urban environments as an incentive to support waste diversion from landfills, decentralize waste separation activities, and increase the transformation of waste materials into valuable raw materials. A case study is discussed for Ethiopia, Ghana, and South Africa with the goal of explaining current processes, urban planning initiatives required for greater implementation, and how they can be interpreted as adaptation initiatives.

**Keywords**

African cities · Urban morphology · Waste management · Methane · Material recycling facilities

## Introduction

### Towards Zero Carbon Cities

Africa has 1.3 billion people (Worldometer 2020) and is the second largest continent after Asia. Many regional studies consider the socio-economic promise of the continent, propelled by urbanization, industrialization, and demographic growth. Achieving this potential would be confronted with a variety of obstacles. Solid waste management is currently one such problem particularly in urban areas. Ten of the fastest growing cities over the next decade are expected to be in Africa. Such cities include Luanda (Angola), Yaoundé (Cameroon), Dar es Salaam (Tanzania), Kumasi (Ghana), Kampala (Uganda), Lusaka (Zambia), Douala (Cameroon), Mbuji-Mayi (Congo), Antananarivo (Madagascar), and Tshwane (South Africa) (Knight Frank 2017). When these cities expand, accumulation of industrial waste becomes expected. Most African cities are generally planned and organized based on colonial urban planning (Home 2015). Little or no town planning has been undertaken since independence. That means little or no concern other than landfills and open dumps built in the 1960s and 1970s for urban waste management.

The main issue is the effect on carbon emissions in all municipal solid waste and cities. The African continent is vulnerable to climate change impacts. Urban areas are critical for effective adaptation to climate change (Revi et al. 2014). More than 50.0% of the world lives in urban areas and these communities partake in industrial practices that generate greenhouse gas (GHG) pollution. Socio-economic inequalities have also seen a dramatic rise in informal settlements in regions that are vulnerable to climate threats, such as natural disasters, public health, damage to homes, and income disruption. Inadequate provision of infrastructure and service delivery are exacerbating risks, especially for low-income households (Revi et al. 2014), and the pace of urbanization has exceeded the ability of many governments to handle it. In 2018, nine African cities pledged to achieve zero carbon cities by 2050. The cities include Accra (Ghana), Addis Ababa (Ethiopia), Dakar (Senegal), Lagos (Nigeria), as well as cities of Cape Town, Durban, Johannesburg, and Tshwane (South Africa). Nairobi (Kenya) and Abidjan (Côte d'Ivoire) are still expected to pledge. Drastic policy, human intervention, and finances will be required to reduce emissions from transport, construction, energy production, and waste management sector (The Economist 2018).

Waste management practices in Africa are not the most effective. Landfill alone contributes between 3.0% and 5.0% of overall industrial emissions (Zhang et al. 2019). Much of the academia and industry focuses on encouraging recycling practices not only as a method of waste disposal from landfills and open dumps, but also as a measure of climate adaptation and mitigation. Africa recycles only 4.0% of its total solid waste (Mohee and Simelane 2015; UNEP 2018). Before landfilling, further exploration of waste collection, separation, and beneficiation is necessary. More Material Recycling Facilities (MRFs) should be considered to boost recycling practices, waste data collection, and alleviate African cities from disposal pressures. Countries such as the United Kingdom have state-of-the-art engineered facilities that divert approximately 50,000 tons of waste per annum from landfills (Ali and Courtenay 2014). In the African context, academic research pertaining to MRF adoption and implementation is far and few in-between while industry or commissioned research is available but restricted between client and supplier. There is a global motivation to dispose of waste from open dumps and landfills as well as to reduce the contribution to emissions. With growth and urbanization proliferating in the coming decade, it is our view that urban planning has a role to play in rising MRFs as a way of mitigating the projected volumes of waste in urban areas.

## Objectives and Chapter Outline

Urbanization and population growth are anticipated to be significant drivers of climate change and pollution in the Global South. Currently, cities accommodate over half of the global population, generate approximately 82.0% of global GDP, account for 70.0% of global energy consumption, and account for over 70.0% of greenhouse gas emissions (Godfrey and Zhao 2016). However, it is estimated that by 2050 global, urban population will exceed 6.7 billion and that nearly 80.0% of population growth will take place in low and middle income countries, where populations are already rising by over one million people per week (C40 Cities Climate Leadership Group 2016). There is therefore an urgent concern to increase climate change mitigation and adaption efforts in African cities. Thus, the main objective of this chapter is to determine how material recycling facilities can be used as tools to facilitate climate change adaptation (and mitigation) in African cities.

To this end, the approach used consisted of exploratory case studies of three waste disposal activities in Ethiopia, Ghana and South Africa, and a review of secondary data, consisting of project reports and scholarly articles. The report is organized as follows: section "Literature Review" addresses the evolution of urban planning in Africa and contextualizes the relationship between climate change and waste management in Africa as well as the rationale for improving waste management and introduces Material Recycling Facilities, how they are planned and designed. The aim is for readers to have an appreciation before going into the case studies. Section "Research Approach" discusses the exploratory case study approach. Section "Case Studies" discusses whether Ethiopia, Ghana and South Africa have urban, waste management, and climate change adaptation plans, strategies, and projects relating to MRFs. Section "Proposed Recommendations and Conclusion" offers climate change and waste management practitioners recommendations and concludes with an assessment of urban planners' roles and responsibilities.

## Literature Review

### Evolution of Sustainable Urban Planning in African Cities

Africa's towns and cities are shaped by laws imported from British colonial rule from the 1920s (Home 2015). Urban Planning, interchangeably referred to as Town and Regional Planning or City Planning, is a discipline within the built environment that is concerned with designing and developing land use (Lubida et al. 2019). It was officially recognized as a profession in the latter twentieth century (Home 2015). Key considerations range from transportation, engineered utilities (water, electricity, sewerage systems), infrastructure, communication, and social amenities (schools, hospitals, parks, economic activities). Spatial and land use planning are the most predominant specializations in the field of urban planning. This is a dynamic activity that requires juggling social, cultural, environmental, and political influences (Lubida et al. 2019). Decisions surrounding zoning, subdivision, and urban planning are taken by land use authorities, in consultation with different stakeholders (Home 2015). Amenities such as waste management have not been deliberate in planning strategies. While there is a global push for cities to become more environmentally responsive and to implement mitigation and adaptation initiatives through policies and strategies – Sustainable Development Goal 13, Target 13.2 – current urban systems are not conducive. A large part of African cities are still centered around colonial town planning principles and regulation, with a morphology characterized by informalities – informal settlement, informal structures, and informal waste pickers.

Postindependence, there was very little evidence of the evolution of land use planning, especially the form that supports sustainability. Throughout the 1980s and 2000, several academics started to re-invent postmodern planning, one driven by economic development and environmentalism (Okeke and Nwachukwu 2019). An increasing body of literature on urban sustainability and innovative planning approaches has begun to shift urban development patterns (Lwasa and Njenja 2012). The term "green technology" was identified as a tool for modern urban planning. This includes deliberate coordination of activities that are energy efficient and promote renewable energy, recycling activities, and more. There are different technologies that contribute towards sustainable urban planning, such as the implementation of enhanced municipal waste management (Laffta and Al-rawi 2018). Four types of eco-technologies that urban planners can start implementing more are:

1. Environmental Technology: a range of technologies relating to waste management, access to water and energy
2. Information Technology: hardware and software coupled with environmental sensor technology to collect environmental data
3. Geographic Information Systems: collect, share, manipulate geographic data and incorporate with Urban Information Systems to guide design and development of land use and environmental plans
4. Communication Technology: software to transfer data and knowledge to different environments

Green technologies are most prevalent and commercialized in Denmark, Sweden, Japan, India, Iraq, Finland, Canada, and the United States of America, just to name a few (Laffta and Al-rawi 2018). Opportunities for African planners to leapfrog are bountiful. Such tools can be used not only for sustainable urban development and waste management but as part of climate mitigation and adaptation strategies. Kenya is not far from realizing the benefits of green technologies. Urban planners are responsible for solid waste management. The National Solid Waste Management Strategy is used to incorporate solid waste management and best practices into urban plans. The plans are then subjected to a Strategic Environmental Assessment, which assesses aspects such as sustainability, proposed waste management programs, as well as relevant services and facilities (Ozoike-Dennis et al. 2019). The authors further state that *solid waste management* (SWM) plans are included *in the Nairobi Integrated Urban* Development Master Plan (NIUPLAN). The SWM plans focus on "improving waste collection and transportation system, closing open landfills, and constructing Material Recovery Facilities, where waste material is to be separated to compost biodegradable material, and recycle non-biodegradable material." In South Africa, municipal departments in charge of solid waste management are required to develop Integrated Waste Management Plans as stipulated in the National Environment Management: Waste Act (59 of 2008). The IWMPs include practical aspects of managing waste in municipalities. It is one of the sector plans that feed into and is guided by the Integrated Development Plans (Sango et al. 2014).

## Climate Change Adaptation

The mean annual temperature in Africa increased by 1.7 °C in the last two decades of the twenty-first century and temperatures are expected to increase faster than the global average in the last two decades (Revi et al. 2014). It negatively impacts on livelihoods, health and sanitation, access to water, agricultural production, and urban infrastructure as well as slows down sustainable development (Reda and Tripathi 2011). Climate change demands new sustainable development strategies through mitigation and adaptation measures. Africa generates just 4.0% of global pollution, but will suffer serious consequences (Reda and Tripathi 2011). Field et al. (2014, p. 1104) define adaptation as a "response strategy to anticipate and cope with impacts that cannot be (or are not) avoided under different scenarios." Adaptation has to do with strategies and plans that can be implemented. They should be linked to development strategies and plans and disaster risk management. By doing this, developmental benefits are identified for the shorter term while reducing vulnerabilities in the longer term (Mimura et al. 2014). The importance of adaptation is influenced by how the issues are outlined (Mimura et al. 2014). They are heterogenous and very few have been monitored and evaluated in great detail. The process of adaptation planning is a complex social process that can create unrealistic expectations. It is done at a national level and implemented at a local level. However, poor coordination as well as recent and quality data can also impede the transition from planning to implementation, as is the case for many developing and developed countries (Mimura et al. 2014).

Accordingly, adaptation responses differ from context to context and involve a multiscale perspective which takes into account cultural, social, environmental, and institutional factors. If climate change is moderate, sustainable development will be greater and if it is high, sustainable development has less impact (Field et al. 2014). It involves designing strategies in the sense of waste management which reduce generation. Given the expected population growth, waste generation is unavoidable. Several primary practitioners interested in adaptation responses are environmental planners as they are planning for physical environments, environments impacted by disasters of climate change (Mimura et al. 2014). They build land use plans enabling or avoiding climate-risk projects (Revi et al. 2014, p. 8). City planning and land management are important (but certainly not the only) resources for policymakers to create climate-resilient city systems and construct capacity for sustainable urban growth at local level, connecting adaptation and mitigation steps, and enhancing livelihoods and quality of life for urban communities. Attention should shift to integrated development planning (Reda and Tripathi 2011). Today, African cities face a critical urban situation marked by a failure of governance structures in the face of increasing waste production combined with strong demographic growth and proliferation of slums and poverty. Strengthening waste management capability is critical for fostering smart urban growth and maintaining cities' democratic, economic, and social stability.

Greenhouse gas emissions from urban waste in developing countries are expected to increase drastically in future (Friedrich 2013) which may result in increased emissions from the waste sector. Increasing recycling, anaerobic digestion, and composting activities can reduce emissions. Waste material is not the only aspect of the value chain contributing towards emissions. Collection and transportation consume approximately 5 $dm^3$ (liters) of diesel per ton of waste, which accounts for 15 kg $CO_2e$ emissions from landfilling alone range from 145 to 1,016 kg $CO_2e$ per ton of wet waste. Aerated and turned windrow composting releases between 172 and 186 kg $CO_2$ per ton of wet waste (Friedrich 2013).

Globally, landfills contribute 3.0% to 5.0% of total global emissions (Zhang et al. 2019) because methane gas is the main carbon emission (Couth and Trois 2012). In Africa, it remains the cheapest and most common disposal method. Landfills are categorized as controlled or uncontrolled dumping or semi/medium/highly engineered facilities. The status of landfill classification structures in sub-Saharan Africa between 2000 and 2018 was evaluated by Idowu et al. (2019). To name a few, they identified 31 landfill sites across South Africa, Botswana, Cameroon, Ghana, Uganda, Kenya, Tanzania and found that 80.0% were listed as 0 and 1 (where 0 was classified as open dumpsites and 1 as controlled tipping). Classifications as non-engineered landfills make it difficult to beneficiate waste and manage environmental impacts such as leachate, groundwater contamination, air borne particulates, and pollution. Vaccari et al. (2019) conducted a technical review of leachate characteristics in landfills and open dump engineering in developing countries. The authors found 21 engineered landfills and 13 dumpsites across Africa as well.

## State of Waste Management in Africa

Local authorities in developing countries are facing significant waste management problems as a result of an increasingly rapid rise in locally generated quantities and the complexification of the forms of waste to be handled. Informal waste collection is also used as an external constraint: it usually takes place under unstable social and environmental circumstances and, a priori, makes waste disposal more complex, managed by a multitude of minor actors. Moreover, data on waste quantities, characteristics, and source is a major impediment particularly for proper planning, implementation, monitoring, and evaluation. Most of the data in this section is based on 2005 and 2012 statistics, which was also used to forecast waste proliferation by various regional bodies, namely, the United Nations Environmental Programme. The daily ratio is the amount of waste generated per capita and per day. Its knowledge is a crucial phase in the introduction of waste management approaches. The amount of waste produced is always very heterogeneous and variable depending on the region, the lifestyles and culture, and the socio-economic level. Changes in lifestyles, living habits, consumption, and demographic growth have a significant effect on the quantity and typology of waste generated. This development often varies from one city to another within the same region, or from one district to another within the same city also according to peoples "living conditions." In sub-Saharan Africa, though per capita waste production rates are lower than in other areas of the world, high population growth, combined with increasing urbanization, would increase unbearably waste generation beyond existing capacities. In this regard, Kaza et al. (2018) forecasts that the overall quantity of waste in Africa will double from 174 million tons per year in 2016 (2.1 billion tons globally) to 269 million tons by 2030 (3.4 billion tons per year in 2025). Africa's waste generation accounts for 9.0% of total global waste generation (Kaza et al. 2018). Developing countries generally have an average annual production of waste per capita ranging from 180 kg to 240 kg, a quantity that is likely to double as a result of industrialization and the evolution of production and consumption patterns if adequate measures are not taken to influence this progression.

*Waste Services and Infrastructure*: Municipalities are responsible for providing waste related services in most African cities (Friedrich 2013; UNEP 2018). Infrastructure includes but not limited collection, transportation, recycling, treatment, disposal, and beneficiation in some cases. However, expensive and insufficient capacity limits their ability to provide sound infrastructure and quality services to all its customers. Lack of payment by customers has a direct impact on municipalities' waste management budget, which restrains their ability to ensure quality services and infrastructure. To deal with this issue, private sector companies and community based organizations partner with municipalities (UNEP 2018). There is a relationship between Willingness to Pay (WTP), affordability, and quality of services rendered.

*Waste Generation*: In 2012 Africa generated an estimated 125 million tons, 65.0% of which was sub-Saharan Africa (80 million tons), and by 2016 it was an estimated 174 million tons. Waste generation is expected to more than double to 244 million tons per annum by 2025 (UNEP 2018) to 269 million tons in 2030 and 516 million tons by 2050 (Kaza et al. 2018). Some of Africa's big economies, comprising of Algeria, Nigeria, South Africa, Ethiopia, and Egypt, are expected to be some of the biggest generators of waste. The forecasted growth is likely to adversely impact the climate if adaptation and mitigation measures are poorly implemented (Fig. 1).

*Waste Composition*: Waste composition has direct implications about how it is collected and disposed. Increasing raising living conditions are correlated with a qualitative shift in waste. As a country's income rises, the biodegradable ratio of the waste often declines as plastic, paper, and other industrial material waste rises (Hoornweg and Bhada-Tata 2012). The environmental effects of such a qualitative mutation are negative as nonbiodegradable waste takes more time to decompose. For Africa, the organic component is still the largest and emits a lot of methane, but it is likely to decline with the economies "industrialization movement." Fifty seven percent (57.0%) of MSW comprises of organic matter (Couth and Trois 2012; Mohee and Simelane 2015; UNEP 2018), followed by plastic (13.0%), paper and cardboard (9.0%), glass and metal (4.0%each), and other material (13.0%). However, composition varies from country to country, depending on economic activities, income level, consumer attitudes, and culture (UNEP 2018) (Fig. 2).

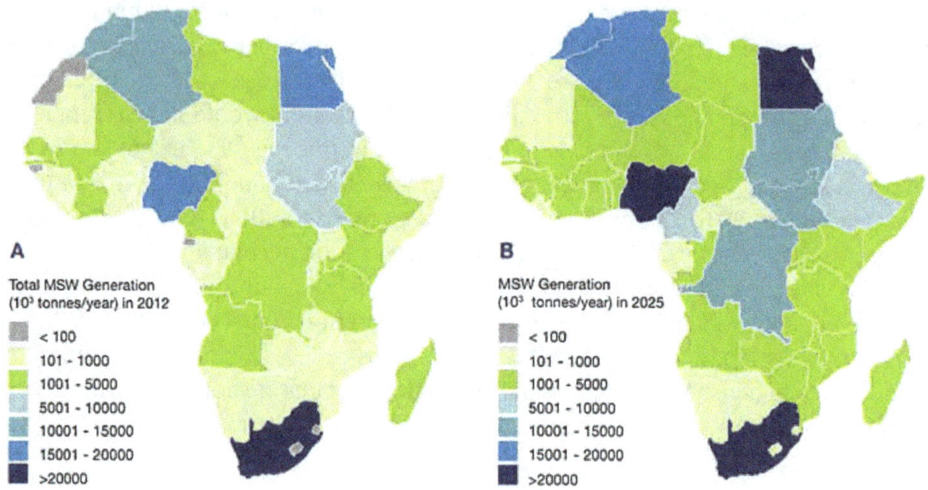


**Fig. 1** Total MSW generation ($10^3$ tons/year) of African countries in 2012 (**a**) and 2025 (**b**)[3] (UNEP 2018)

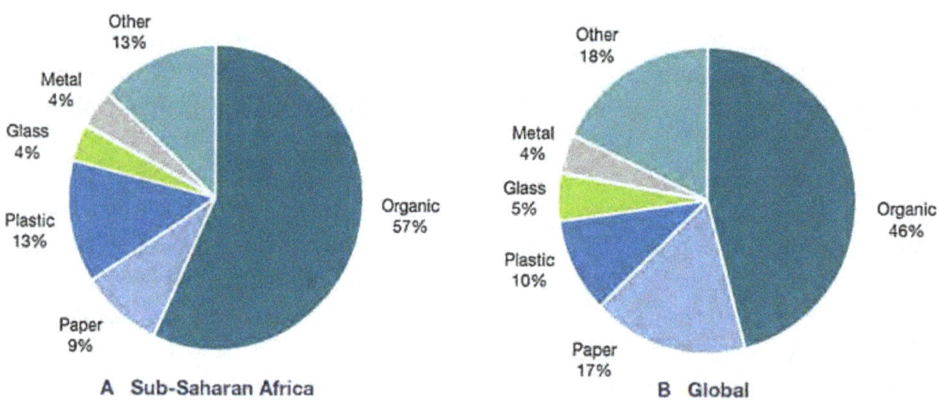


**Fig. 2** Municipal solid waste composition (UNEP 2018)

*Waste Collection*: Uncollected waste remains very large in Africa. It might be necessary to note, however, that certain African cities have adopted improved management selection techniques. Some work has shown a spike in private-sector collections (UNEP 2018). Choosing a treatment method entails multiple economic, social, and environmental factors such as waste generation, population density, and location of waste collection centers (Mohee and Simelane 2015). Unregulated landfills and informal recycling are the main waste disposal methods in African countries and there is a high demand for waste services and low supply. Majority of the budget for MSW in developing countries is spent on waste collection yet only 55.0% of total waste generated was collected. Although waste collection is expected to increase to 69.0% by 2025 (UNEP 2018), where it will be disposed remains a concern. Collection services are most present in urban areas compared to suburban and rural areas. A common practice is door-to-door collection, by both public and private sector contractors. In peri-urban areas, community-based organizations are very important.

*Disposal Methods:* Landfilling and open dumping are the most common waste practice on the continent because both are considered the cheapest forms of disposal (UNEP 2018). Landfills are categorized differently, but the most common categories are controlled and uncontrolled. The nature of landfills and open dumps have little to no consideration for human health and the environment. Poor sanitary engineering and waste treatment results in groundwater contamination, leachate overflow, and an unpleasant stench. Across most landfills and open dumps in Africa, scavenging activities are inevitable (Idowu et al. 2019). Households near disposal sites are exposed not only to waste-borne diseases, but sharp objects and dangerous chemical excretions. Furthermore, GHG emissions are exacerbated due to poor waste minimization prior to disposal and open burning (9.0%). As the sites become full, the disposed waste is burned in order to reduce volumes and create more space (Fig. 3).

*Opportunities for waste recycling*: Most literature about recycling acknowledges the opportunities presented by such activities. In the African context, a lack of empirical data is one of the reasons why it is not a common practice particularly within the public sector. Recycling is not a well-known method in most municipalities, and this may be attributed to a lack of knowledge. Recycling activities are done by private contractors and informal waste pickers and they represent only 4.0% of total recycling activities on the continent (UNEP 2018). Although these are positive strides, waste material from developed countries continues to be dumped and lead to new waste streams, and most municipalities are not well equipped with the necessary logistics required for handling recycling. The informal sector, which comprises of mainly waste pickers, largely practices recycling for economic reasons. They play an important part towards reducing

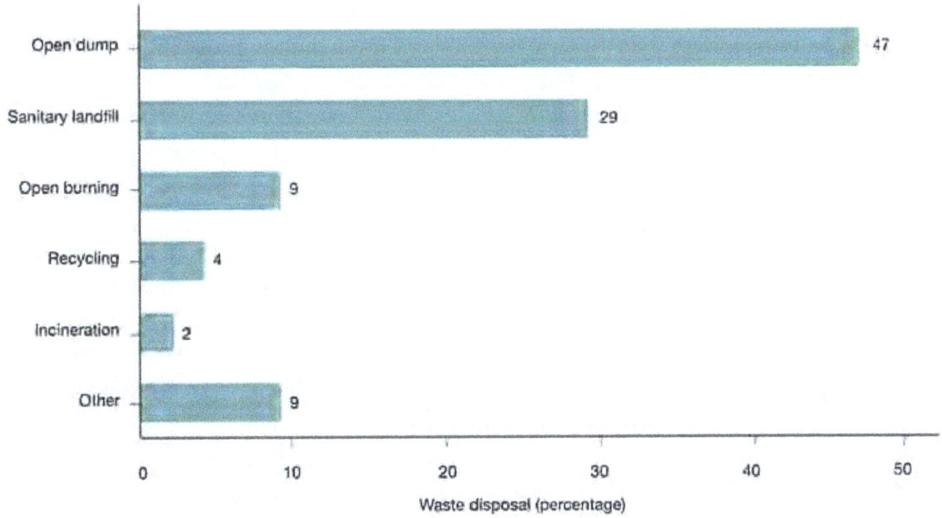


**Fig. 3** Methods of end-of-life MSW disposal in Africa (UNEP 2018)

carbon emissions (Nzeadibe 2013; Oduro-Appiah et al. 2019). Recyclables such as plastic, glass, paper, and metals are supplied to commercial businesses. Organic waste which releases the most emissions due to its $CH^4$ composition is seldom recovered for large-scale impactful composting. Waste pickers in South Africa save municipalities an average of US$30 – US$70 million annually in landfill airspace, with little to no financial or operational support (Godfrey et al. 2016).

## Material Recycling Facilities

There are many alternative waste management options available but not all are economically or contextually viable for the continent. Recycling activities and energy recovery technologies are currently being implemented but both on a very small scale. Technologies that can be reviewed are biological treatment, mechanical biological treatment, thermal treatment (incineration, gasification, and pyrolysis), and material recycling facilities (UNEP 2018). As mentioned in previous sections, Africa's recycling rate is only 4.0% annually. Municipalities are impeded by financial and infrastructure costs and private sector recyclers' business models do not cater to all waste producers due to affordability. Informal sector recyclers are an important aspect of the waste value chain (Godfrey et al. 2016). Material recycling facilities are an integral part of an integrated waste management model and offer a great opportunity for all the role-players to converge. The study focuses on material recycling facilities due to their ability to be implemented immediately, contextually, and cost-effectively.

*Defining MRFs:* An MRF is a plant where waste is received, sorted, and processed for revenue generation and to reduce negative environmental externalities (Hosansky 2018; Zafar 2019). The plants can be classified either as dirty or clean. Dirty MRFs process mixed waste which requires a lot of manual labor while clean MRFs tend to process waste that was separated at source and reduces waste contamination. MRF sizes vary from context to context. Zafar (2019) states that small MRFs process less than 10 tons per day and price ranges between US$500,000 and US$1 million. Price determinants are costs of building material, location, and the level of automation required. Large MRFs process more than 100 tons per day and are usually located in large cities. Due to the high automation process, the budget runs into millions of dollars. In a study by South Africa's Department of Environmental Affairs (2019), large MRFs can process between 1,200 and 1,500 tons of waste per month while small MRFs can process between 300 and 500 tons per month. Despite the varying sizes, MRFs have an important role within the waste value chain, primarily to improve waste separation and reduce waste streams before implementing additional management methods – see Fig. 4.

*Role-players and Decision-Makers*: there are many role-players along the waste value chain and their level of involvement is based on different contexts. Traditional actors include municipalities, private sector (business and industry), the informal market, waste producers, and civil society. Effective or noneffective coordination determines the level of successful implementation. During the planning stages, different levels of planning and decision-making constituencies in municipalities

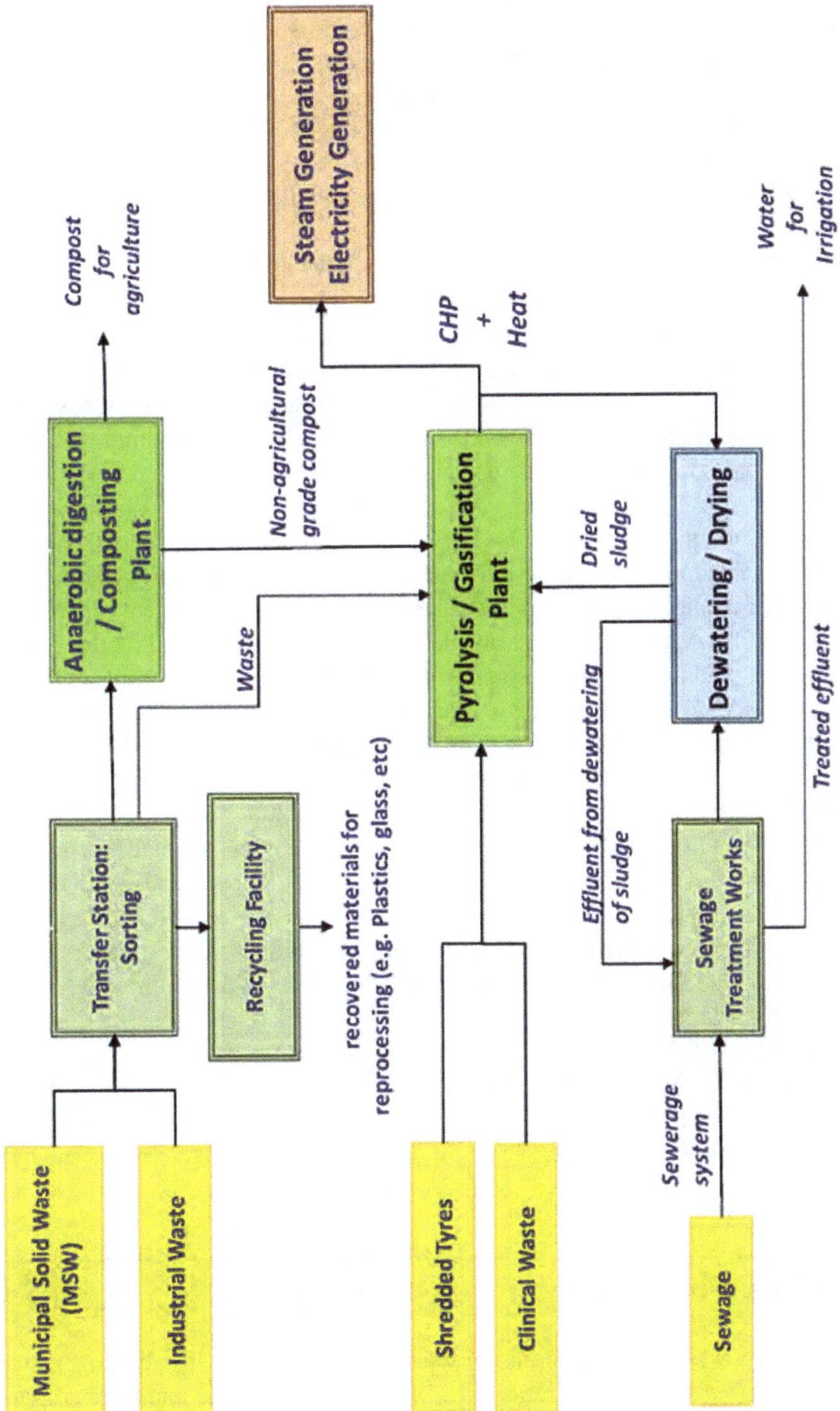


**Fig. 4** Integrated waste treatment system (Trois 2014)

are involved. They need to be aligned to ensure sustainable waste management is implemented. According to Sango et al. (2014, p. 223) "*stakeholders in solid waste management include, but are not limited to, solid waste managers, technical/utility managers, municipal managers, chief financial officers, mayors, councillors and the public at the local municipal level.*" Their lack of involvement can negatively or positively impact technical design due to incorrect data, poor policy, and regulation considerations and accurate budgeting for building material and equipment. This is seen in the growing body of literature encouraging informal waste pickers to be integrated into solid waste management strategies.

*Technical Design Considerations:* Good knowledge and data on waste quantities and characteristics are imperative to an MRF's efficiency. However, 100.0% material recovery is not guaranteed due to the quality of materials and sorting efficiency as well as market conditions. Figure 5 presents a technical design for a proposed dirty MRF in South Africa. When waste arrives at the plant – either in waste bags or trucks – it is taken for preliminary treatment where waste volumes are reduced. Equipment such as bag splitters is used together with manual laborers to tear waste bags in preparation for sorting. At this stage, recyclables are sorted depending on the waste characterization (paper, plastic, glass, and metal). The quantities determine whether sorting will be entirely manual, semi-automated, or fully automated. Manual sorting is still important particularly for quality control purposes (Cioca et al. 2018; Department of Environmental Affairs 2019). The recyclables are then compacted and prepared for the market, for composting, and anaerobic digestion or residual is taken to the landfill.

*Financing and Economic Instruments*: It is important to standardize what a big and small MRF design will look like. The design should consider waste generation quantities and offtake markets for recyclables, supply and demand principles. Although mechanical MRFs are considered more economic because they can process 3.0 tons per hour with approximately seven operators compared to manual MRFs that process 1.0 ton per hour with 12 operators (Cioca et al. 2018), labor intensive activities should be highly considered in the African context because of job creation opportunities. Financing models range from public-private partnership, to fully publicly owned and operated. The proposed model is a public-private partnership where risks are shared and efficiency is maximized (Department of Environmental Affairs 2019). Economic instruments such as subsidies and taxes are designed to change consumer behavior. In the case of waste, they aim to reduce waste generation or induce waste diversion from landfills and dumps to recovery and recycling activities. Furthermore, they can also be used as a source of revenue for municipalities and implement service delivery (Nahman and Godfrey 2010).

## Research Approach

An exploratory case study research approach was used. This involved empirical investigation of current consideration of MRFs in urban, waste management, and climate adaptation plans and strategies in selected African countries. The exploratory approach is intended to provide an understanding of what is available and set

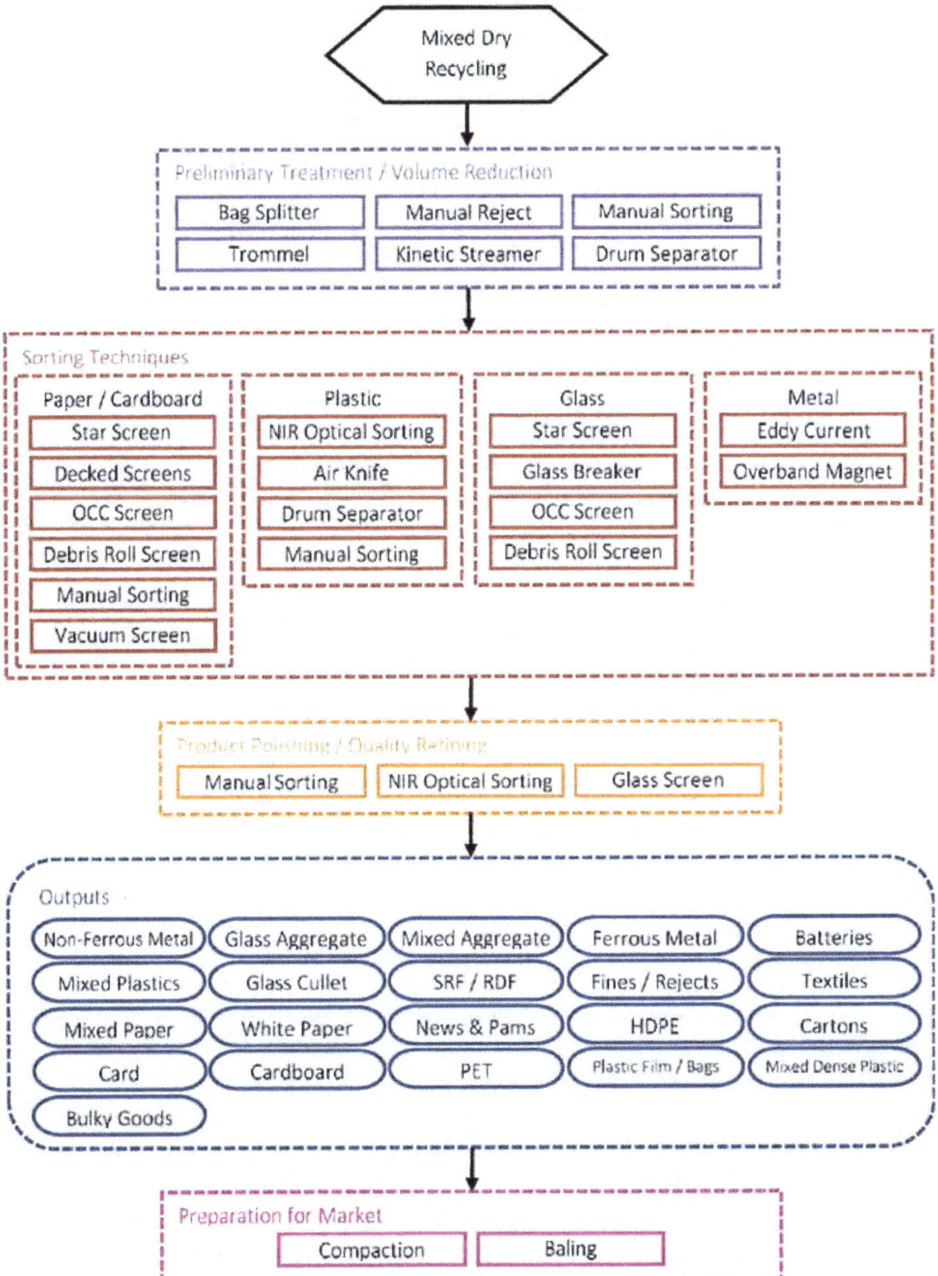


**Fig. 5** Material recycling facility process (Department of Environmental Affairs 2019)

precedent for future investigation with participants and be more descriptive. Ethiopia, Ghana and South Africa were identified as targeted case studies based on three primary criteria: countries with the fastest growing cities in the respective regions, cities that committed to become zero carbon by 2050, and cities with some data on recycling, landfills, and open dumps – see Table 1. Selecting a case study

**Table 1** Case study selection criteria of cities

| Fastest growing | Committed to be carbon free | With landfills and open dumps |
|---|---|---|
| Angola (Luanda) | *Ethiopia (Addis Ababa)** | Botswana |
| Cameroon (Douala) | *Ghana (Accra)** | Cameroon |
| Cameroon (Yaonde) | Ivory Coast (Abidjan) | Dem.Rep of Congo |
| Dem.Rep of Congo (Mbuyi-Mayi) | Kenya (Nairobi) | *Ethiopia** |
| *Ghana (Kumasi)** | Nigeria (Lagos) | Gambia |
| Madagascar (Antananarivo) | Senegal (Dakar) | *Ghana** |
| *South Africa (Tshwane)** | South Africa (Cape Town) | Guinea Bissau |
| Tanzania (Dar es Salaam) | South Africa (Durban) | Kenya |
| Uganda (Kampala) | *South Africa (Johannesburg)** | Namibia |
| Zambia (Lusaka) | South Africa (Tshwane) | Nigeria |
| | | Sierra Leone |
| | | *South Africa** |
| | | Tanzania |
| | | Uganda |
| | | Zimbabwe |

**KEY**: * = countries with the fastest growing cities that are committed to being carbon free and have landfills/open dumps

per region allows for a comparative exploration. Secondary and grey literature provided further support and rationale for selecting these three countries. Two challenges experienced with Francophone and Lusophone countries were firstly, limited availability of recent data and secondly, some available data is presented in French and Portuguese.

## Case Studies

Based on the range of impediments identified in section "Literature Review," the case studies explore whether the identified countries have urban, waste management, and climate change adaptation plans, strategies, and projects relating to MRFs. Discussions are themed according to factors used to determine suitable MRF design and adaptation planning, such as (1) Urbanization Dynamics, (2) Urban Waste Impediments, (3) Current and Proposed Strategies, (4) Role of Informal Waste Collectors, as well as (5) Nationally Determined Contributions (NDCs) focusing on waste management technologies.

### Urbanization Dynamics

As in many African countries, rapid urbanization in Ethiopia, Ghana, and South Africa is a result of increasing economic activities particularly in large cities. Addis

Ababa is the capital city and seat of the Federal Republic of Ethiopia. It has a current population of 3.3 million (23.8% of all urban dwellers) and extends across 540 $km^2$ of land space. The administration is known as a City Government and it has 10 subcities known as *Kebele*. In Western Africa, the City of Accra – which is part of the Greater Accra Metropolitan Assembly – is the capital city of Ghana and accounts for 20.0% (1.5 million) of the country's total population (The World Bank 2017). More than 53.0% of Ghana's total population currently reside in urban areas and this figure is projected to reach 65.0% by 2030. The Accra Metropolitan Assembly (AMA) is responsible for all local government activities within Greater Accra. Further down south, the City of Johannesburg (CoJ) is the largest city in South Africa. With a population of 5.6 million (66.4%) spanning across 11 regions and 1,645 $km^2$ of land space. Like its counterparts, the City of Johannesburg Metropolitan Municipality is responsible for service delivery. Provision of basic amenities and infrastructure has not kept up with the urbanization rates in all three cities, thus resulting in more urban tensions such as social unrest and sprawling which induce climate change impacts.

## Urban Waste Impediments

Waste management and poor sanitation infrastructure are a few of the services that governments continue to grapple with especially in low income areas. In 2013 Accra was generating 2,200 metric tons of waste per day and approximately 1,500 to 1,800 tons was collected by private sector. Between 2017 and 2018, generated quantities increased to 3,000 tons per day (Accra Metropolitan Assembly 2019). Waste collection and transportation alone cost the metropolitan GHC 450,000 (approximately US$77,000) per month and an additional GHC 240,000 (approximately US$41,000) to maintain dumpsites, resulting in 58.0% of waste and sanitation budget being spent on collection and transportation (Accra Metropolitan Assembly 2019). One of the factors that made collection and transportation difficult was the city's spatial planning. Oteng-Ababio et al. (2013) stated that "*Accra did not have the advantage of being a wholly planned city, since spatial planning was introduced after the town had developed on its own. Spatial planning is inconsistent at best, with semblances of colonial-era planning in (some) indigene areas such as Old Accra and Korle Gonno in Ga Mashie while migrant communities such as Nima, Sabon Zongo and Old Fadama continue to grow unplanned. Such unplanned settlements tend to be classified as 'poverty pockets' in the city, with a history of poor municipal investment in infrastructure services and under- and unemployment (particularly among the youth*)" (p. 98), resulting in lasting impacts on sanitary service delivery. Tuani (2008) resonates with the authors by stating that the unplanned nature of the city increased waste management issues, in particular land acquisition for landfills/open dumps and waste disposal sites. Additional challenges include poor separation at source, limited recent data, and limited optimization of other disposal plants.

Similarly, in Addis Ababa, waste collection and disposal are poorly managed. Citizens generate 730,000 metric tons of waste per annum (between 2,200 and 3,000 tons per day). Koshe landfill is the only landfill and biggest one in the city. It is located 13 km in the outskirts of Addis Ababa and the travel distances from other parts of the city make it inaccessible and costly. This results in poor collection and disposal in the nearest area. Just over 65.9% is collected by the metropolitan municipality and the remainder is disposed in unauthorized areas such as open fields, streets, ditches, and any available space within the city. Citizens do not have conviction in the metropolitan's ability to deliver waste management services, resulting in limited to no cooperation with waste related operations. Urban development and sprawling are occurring at a rate that government cannot service all households (Geda et al. 2011). Poor urban and road infrastructure make it difficult to collect data in highly dense settlements, further exacerbating effective implementation. The same sentiments are shared by Johannesburg where residents generate 1,492,000 tons of waste per annum and the four landfills cannot continue taking (City of Johannesburg Municipality 2011). Although other disposal measures such as recycling and composting are part of the city's management strategy, effective implementation is challenged by urbanization dynamics. In 2015, the Department of Environmental Affairs estimated that Johannesburg had between 5 and 8-years landfill airspace remaining (Department of Environmental Affairs 2018). Fast track to 2020, at least 5 years are up and pressure is increasing with population growth.

## Current and Proposed Strategies

In Accra, AMA introduced two collection methods to help streamline collection and transportation: Communal Container Collection (CCC) and HtH (house-to-house). Households were required to pay a monthly collection fee, but those residing in low-income areas were daily and weekly wage-earners, they could not afford the service. Furthermore, areas where CCCs were located had bad road infrastructure. This presented locality and accessibility challenges. Since then there has been an increase in private sector participation, particularly waste collection. Affordability remains an impediment. Because waste was poorly collected, it was difficult for AMA to promote source separation. The ripple effect was poor planning and implementation of other disposal methods like composting, digestion, and recycling. Almost a decade later, waste management officials cite (1) community awareness, (2) effective collection systems, (3) limited disposal capacity, and (4) unplanned nature as ongoing challenges. Figure 6 discusses an investigation of Accra's urban planning, waste management and climate change plans and strategies to ascertain consideration for MRFs as an adaptation tool.

In the far East, containers are placed in common places each Kebele, close to main roads where waste trucks can access. Pushcarts are used to collect waste from each household at a fee. Those who cannot afford the fees carry their own waste to the container. Services are provided by Kebele workers and street sweepers (Cheru 2016). In 2017, the Reppie Waste-to-Energy Facility was commissioned, producing 25 MW of

**Plans and Strategies - Accra**

Accra Spatial Development Framework (2017) recognizes the various challenges presented by poor waste collection and disposal propelled by insufficient infrastructure and urbanization. It proposes that AMA 'explores job creation and economic opportunities associated with engineering service delivery, for example decentralize waste collection to locals / individuals. Furthermore, preparation and implementation of sanitation action plans for all leading urban centers, including related statutory regulations and by-laws ensuring effective collection, disposal and treatment of solid, liquid and toxic waste. The ASDF also proposes New Solid Waste Management Schemes and Environmental and Land-Use Policies but is not explicit about disposal methods, location determination or MRFs. Although it states that district level climate change mitigation and adaption strategies must be developed in order to plan for and implement requirements for climate change, additional details are not provided. However, detailed consideration is given in the Ghana National Climate Policy (2013). The policy recognizes the opportunities presented by beneficiating waste, problems with land acquisition for waste disposal as well as the need to increase recycling activities as means to reduce carbon emissions. It is reported that in April 2020, AMA entered into a public-private partnership (PPP) for Waste Transfer Site and Material Recovery Facility (MRF). Zoomlion currently operates a 200-tonne recycling plant in Accra. No reference of MRFs in Integrated Urban Environmental Sanitation Master Plan (IUESMP), Metropolitan Land Use and Transport Plan for Greater Accra Region, Emergency Solid Waste Management Improvement Programme13 (E-SWMIP), the Accra Climate Strategy and Integrated Urban Environmental Sanitation Strategy and Master Plan.

**Fig. 6** Investigation and discussion of MRF consideration in Accra. (Source: Author's own compilation)

electricity. The facility is located next to Koshe Landfill site and waste has since been diverted for incineration instead, thus reducing emissions. There are traces of recycling activities but on a small scale at disposal sites (Geda et al. 2011). Private sector, community-based organizations, informal waste pickers, and international donor agencies are the main participants promoting some levels of at source separation and income generating activities. IGNIS was a joint project between Ethiopian and German partners. The project started in 2014, with the aim of implementing *income generation and climate protection by valorizing municipal solid waste in a sustainable way in emerging mega cities*. Strategies included creating individual waste recycling projects that could create employment and income opportunities while reducing GHG emissions. The project lasted until 2018 and provided valuable data to waste management agencies, namely composting (Gerji and Ataklet Tera Composting pilots) and biogas (Sidest Kilo and Menelik II preparatory School pilot) (Cheru 2016). Figure 7 also investigates Addis Ababa's urban planning, waste management, and climate change plans and strategies to ascertain consideration for MRFs for adaptation.

Johannesburg has better waste management infrastructure and initiatives. Pikitup is CoJ's waste management agency responsible for collection and disposal across 11 regions. As with many cities on the continent, increasing urbanization trends are causing stain on quality service delivery. CoJ has numerous private-public-partnerships throughout the waste value-chain to reduce the high costs of collection and disposal incurred. Most PPPs fall within waste collection and disposal. Formal urban areas have scheduled collection dates and residents have municipal bins. Some informal areas do have some sort of waste service and others do not. This results in pockets of illegal dumping. There are four landfill sites in different ends of Johannesburg and all four have reached their airspace capacity. Although CoJ has championed waste diversion activities such as recycling, composting, and pilot projects for small scale biogas projects, there is a need to increase these activities. The recycling sector is relatively well established compared to Accra and Addis Ababa. Associations such as South African Plastics Recycling Organization, National Recycling Forum, and the Institute

**Plans and Strategies – Addis Ababa**

Addis Ababa City Structure Plan (2017-2027) shares strategic plans and land use tools to improve urban development in the city. It recognises waste and climate related challenges as well as the need for improved recycling activities. The CSP states that it is preferable to make use of a mix of centralized and decentralized offsite and onsite liquid waste management systems as well as wastewater treatment technologies that do not consume large tracts of land. Interestingly, the plan proposes that three transfer stations with material recovery facilities (south, east, west) should be developed at each waste transfer station. The Draft Work Plan for Addis Ababa - Climate and Clean Air Coalition, prepared by the U.S. Environmental Protection Agency (2015) addresses the direct relationship between climate change and improved waste management.

The World Bank has a number of urban and climate change related projects underway in Addis Ababa. 'Enhancing Urban resilience Programme' was launched in 2013 to help cities strengthen their ability to prepare for and adapt to changing conditions, and to withstand and recover rapidly from disruptions related to climate change, natural disasters, and other shocks and stresses. The programme serves as an umbrella for delivering the analysis, rationale, and support local governments need to make resilience part of their urban management agendas. The 'Urban Local Government Development Project' (2008-2014) was designed to financially support the Ethiopian government "to support improved performance in the planning, delivery, and sustained provision of priority municipal services and infrastructure by urban local governments", inclusive of waste infrastructure.

**Fig. 7** Investigation and discussion of MRF consideration in Addis Ababa. (Source: Author's own compilation)

**Plans and Strategies – Johannesburg**

Various national policies require municipalities to develop Integrated Waste Management Plans (IWMPs) and Integrated Development Plans (IDPs) every five years. The IWMP recognizes that development planning in the City must also consider waste service provision and therefore early engagement and communication between planners and those involved in waste management is needed. Composting facilities, MRFs and buy back centers are noted as action plans for waste diversion. Similarly, the IDP 2018/19 acknowledges the need to improve waste management infrastructure supportive of waste minimization and recycling as well as climate change. CoJ has Environmental Sustainability Strategy and Action Plan (2019) encourages spatial transformation in cities. This includes applying existing and develop municipal planning instruments to integrate and secure ecological infrastructure into the urban fabric.

Tedcor piloted a materials recovery facility (MRF) in Wadeville, near Germiston in Gauteng, in partnership with local buy-back centers. Droplets for Life – a private waste management company - have their own MRF situated in Wadeville, Gauteng. They receive waste through this facility from clients and from their our own waste. The facility processes approximately between 25 to 30 tons of recycling per month and employed 12 people currently from the disadvantage background. Training and up-skilling were provided to them according to their job specification. In 2014 CoJ commissioned a dirty MRF pilot at the Robinson Deep landfill in Turffontein, Johannesburg, Gauteng under the 'Separation at Source Programme'. The facility was processing 4,970 tons per annum of recyclable materials in 2014. MRFs do not tend to go beyond pilot stage due to low rate of recyclable recovery (Simelane, 2016)

**Fig. 8** Investigation and discussion of MRF consideration in Johannesburg. (Source: Authors own compilation)

of Waste Management of Southern Africa collaborate with private sector, civil society, and academia to promote greater adoption. As with this Figs. 6, 7, and 8 discusses MRF consideration as an adaptation tool for Johannesburg.

## Role of Informal Waste Collectors

Informal waste collectors play an integral role along the waste value chain. Through their activities, waste collectors assist municipalities to collect, separate, and recycle waste material because they do not have sufficient financial and nonfinancial capacity to provide efficient waste management alone. They operate in similar approaches despite not being recognized or integrated in municipalities' strategies and plans. Due to socio-economic factors, waste collectors continue to operate alongside the municipalities, collecting waste from household to household, designated communal containers and institutions.

In Accra, informal waste collectors continue alongside AMA but are not officially recognized. Known as Kaya Bola, they typically port waste from households or markets for a fee, separate the various material, and sell the recyclables for an income. Not being legitimized by local government induces ill-treatment and hostile social environment that keep them operating in an unorganized/informal manner (Oteng-Ababio et al. 2013). Presently, Kaya Bola collect and sort 25.0% of total waste generated for income but dispose worthless material in open drains and streams. The only sanitary landfill known as Knope has reached its capacity, leaving a few operational dumpsites scattered across the region, namely, Oblogo and Sarbah (Tuani 2008; Baah and Kharlamova 2018). It is located 37.0 km outside Accra which makes collection and transportation fees very high (The World Bank 2017). The distance and nature of road infrastructure from waste producers to Knope or other dumpsites is too far for informal waste pickers, more so for the small transport modes they use, for example, motorkings (Accra Metropolitan Assembly 2019).

Presently, informal waste collectors are not recognized as part of Addis Ababa's integrated waste management system. They are seen more as a potential enemy rather than partner (Cheru 2016). Waste pickers collect and sort for valuable recyclables and take them to sell at the Merkato Market place, the biggest market in Ethiopia. Various buyers are located there, ranging from independent buyers to factories. In Johannesburg, the lack of landfill space is increasing pressure on the municipality to integrate informal waste collectors along the waste value chain. The Wastepickers Empowerment Programme was launched in 2011 with the aim of creating a database of waste collectors in the metropolitan, skills training, assistance to establish cooperatives, provide protective clothing, and assist with fund raising. A pilot program was launched in 2014 and more than 500 waste collectors were registered and benefited from various skills training and personal protective equipment whiled 55 cooperatives participated in separation at source (Baker et al. 2016). The program can be scaled to other regions in Johannesburg but some lessons from the pilot would need to be considered:

- Consultation with different municipal stakeholders and recycling associations
  - Identifying suitable sites for sorting and storage
  - Establishment of guidelines for waste collectors
- Identification of waste collectors
- Sourcing external funding for training and equipment

There are opportunities to still officially integrate informal waste collectors in the city's strategies and plans. Associations such as the African Reclaimers Organization represent collectors' views in this regard. The Department of Environmental Affairs estimates that there are over 62,000 waste collectors in the country, with 40.6% operating as trolley-pushers (Mathye 2019).

Overall, various proposals and plans drafted by international organizations such as The World Bank and the African Development Bank stress the need to integrate informal waste pickers into waste management and increase waste transfer stations as strategies for climate adaptation. Their involvement in the waste value chain would boost collection rates and income.

## Nationally Determined Contributions

The Paris Agreement requires every country to produce Nationally Determined Contributions (NCD). These are climate change policies detailing each country's plans towards adaptation and mitigation. Ethiopia, Ghana, and South Africa developed their NCDs. Ethiopia's plan recognizes carbon dioxide ($CO_2$), methane ($CH_4$), and nitrous oxide ($N_2O$) as priority gases that need to be mitigated by 2030. Key sectors producing these gases are agriculture, forestry, transport, industry, as well as waste and cities. In its medium to long term actions, drought relief is an action point where the government will "ensure the uninterrupted availability of water services in urban areas to make them comfortably and productively habitable irrespective of droughts through planning and construction of dams or deep wells, deployment of water saving technologies, and wastewater treatment infrastructure." South Africa's NCDs target agriculture, land use, and waste as targeted sectors for GHG reductions. However, energy is the primary focus.

Lastly, Ghana's adaptation and mitigation commitments strongly feature solid waste management in urban areas while ensuring positive socio-economic derivatives. The country committed to adopting alternative urban solid waste management by 2030. As an adaptation policy action, Ghana committed to building standards for urban waste management through city-wide infrastructure planning. Three mitigation action points include:

- Improve effectiveness of urban solid collection from 70.0% to 90.0% by 2030 and dispose all to an engineered landfill for phase-out methane recovery from 40.0% in 2025 to 65.0% by 2030.
- Scale up 200 institutional biogas in senior high schools and prisons nationwide.
- Double the current waste to compost installed capacity of 180,000 ton/annum by 2030.

There is an opportunity for MRFs to feature as adaptation or mitigation actions depending on each country's resources, more so for Ethiopia and South Africa.

## Proposed Recommendations and Conclusion

It is evident from current strategies and plans in all three cities that there are opportunities for improved MRF adoption as well as consideration. Urban planners are responsible for designing and developing land, specializing in spatial and land use planning traditionally. Their involvement was seen in activities, decisions, strategies, and plans for infrastructure development and service provision which are prepared at local government, as seen in Accra's Spatial Development Framework, City of Johannesburg's Integrated Waste Management, and Addis Ababa's Urban Renewal Initiative. As part of the planning process, waste management and/or climate change officials could be included in the process. This was evident in the case of Kenya.

It was also evident that there is a need for the planning process to evolve with the profession and embrace the global push for sustainability and climate change by being more cognizant about waste management and climate change. A central point where all three disciplines converge is land: access to suitable land for disposal, proximity of land to citizens, and possibilities of land inducing negative or positive climate change impacts. In addition, knowledge and implementation of MRFs is generally unknown, with the exception of South Africa, where the Department of Environmental Affairs have taken interest. Climate change adaptation already

**Table 2** MRFs as a climate adaptation strategy

| Challenge | Recommendation | Adaptation strategy |
|---|---|---|
| Local governments spend at least 50% of total waste budget on waste collection and disposal | Consider exploring small-scale decentralized MRFs to address waste collection, transportation, logistics as well as improve data collection and beneficiation opportunities – climate adaptation strategy | Include more waste management and climate change matters in urban plans<br>Integrate informal waste pickers in city waste management plans to create job in the informal sector |
| Waste generation and quantities are exacerbated by urbanization and population growth, straining local governments' ability to provide services | Using plans and strategies, planners identify land closer to waste producers and socio-economic activities driving waste generation | Use Green Technology such as GIS and IT to map generation hot spots, determine material flows and data |
| | Decentralize MRFs on a smaller scale to improve access. Closer to informal communities, shorter distance to travel and potentially more material recovery | Design MRFs guided by technical and social dynamics around identified site |
| Informal waste pickers are not recognized by local governments despite positive impact in the waste sector | Decentralized small MRFs can assist to streamline waste pickers involvement<br>Create associations or cooperatives for informal waste pickers | Provide training and technologies to women and people in informal communities to enable recycling of waste into other marketable products (e.g., paper briquettes, etc.) |
| | | Develop zones in cities, suburbs and settlements to streamline the collection of waste by informal waste pickers |

In Africa, where most countries are classified as low- and lower-middle income, there are many parameters to consider when selecting AWT technologies, including:

- sensitivity to the quantity and quality of waste;
- capital investment costs
- operation and maintenance (O&M) costs
- cost recovery potential
- efficiency of the technology
- end-product utility
- environmental impacts of operation
- land, energy and water requirements
- availability of local competences to properly manage the system
- inclusion of both formal and informal actors in the system

**Guidance note: Lessons learned**

- An appropriate technology should not be considered in isolation; it must be placed within an integrated waste management strategy.
- Both formal and informal actors must be considered to insure the efficiency of the waste management strategy (high level of waste collection, improvement of recycling and/or reuse, high material and energy recovery, social inclusion and income generation for the poorest people).
- The choice of the appropriate technology should consider the following:
  - The technology should be proven based on effective operating experience in African/low-income country contexts.
  - The technology should be economically viable in the context of low-income countries.
  - Technical issues related to the implementation of the technology should be easy to fix: the presence of qualified human resources for management and maintenance must be considered.
  - Low cost technologies are preferred as waste treatment solutions.
  - The technology should be an opportunity for social innovation (job creation, social inclusion, income generation).

**Fig. 9** Measures to determine appropriate waste technologies (UNEP 2018)

involves plans and strategies, thus once again there is an opportunity to improve adoption and implementation of MRFs as a land-use and adaptation strategy, and urban planners can aid to drive that inclusion. Table 2 below suggests how including MRFs can be considered as a climate change adaptation measure for the three cities and beyond:

*For MRF Facility Design:* it is important to determine and measure the appropriateness of waste technologies for the continent because of varying drivers, pressures, and parameters (Fig. 9).

Although Africa contributes 4.0% of the worlds' GHG emissions, with these proposed adaptation strategies, many cities can reduce future emissions and leapfrog into full sustainability despite the current urban and waste related challenges.

## References


Accra Metropolitan Assembly (2019) Accra Resilience Strategy

Ali M, Courtenay P (2014) Evaluating the progress of the UK's material recycling facilities: a mini review. Waste Manag Res 32:1149–1157. https://doi.org/10.1177/0734242X14554645

Baah D, Kharlamova M (2018) The impact of landfill sites in Accra-Ghana on the surrounding environment. Peoples's Friendship University of Russia, Moscow

Baker M, Mamela S, Rampete N (2016) Empowerment of waste pickers in the City of Johannesburg, creating an enabling environment for the informal waste sector to participate in municipal

waste service delivery. In: Proceedings of the 23rd WasteCon Conference 17–21 October 2016, Emperors Palace, Johannesburg, South Africa, p 7
C40 Cities Climate Leadership Group (2016) New perspectives on climate finance for cities: finance solutions for new and emerging infrastructure approaches to urban climate mitigation and adaptation. Siemens and Citi, London
Cheru M (2016) Solid waste management in Addis Ababa – a new approach to improving the waste management system. Metropolia University of Applied Sciences, Helsinki
Cioca L, Ferronato N, Viotti P et al (2018) Risk assessment in a materials recycling facility: perspectives for reducing operational issues. Resources 7:85. https://doi.org/10.3390/resources7040085
City of Johannesburg Municipality (2011) City of Johannesburg integrated waste management plan. City of Johannesburg, Johannesburg
Couth R, Trois C (2012) Sustainable waste management in Africa through CDM projects. Waste Manag 32:2115–2125. https://doi.org/10.1016/j.wasman.2012.02.022
Department of Environmental Affairs (2018) South Africa state of waste – a report on the state of environment. First draft. Department of Environmental Affairs, Pretoria
Department of Environmental Affairs (2019) Design guide for material recovery facilities (MRF) in South Africa. Department of Environmental Affairs, Pretoria
Field CB, Barros VR, Dokken DJ et al (2014) Climate-resilient pathways: adaptation, mitigation, and sustainable development. In: Climate change 2014 impacts, adaptation, and vulnerability. Cambridge University Press, Cambridge, pp 1101–1131
Friedrich E (2013) An investigation into the emissions of greenhouse gases associated with the disposal of solid waste in the eThekwini municipality. University of KwaZulu-Natal, Durban
Geda N, Sundaraa R, Seboka B (2011) Challenges and opportunities in municipal solid waste management: the case of Addis Ababa City, Central Ethiopia. J Hum Ecol 33:179–190. https://doi.org/10.1080/09709274.2011.11906358
Godfrey N, Zhao X (2016) Financing the urban transition for sustainable development: better finance for better cities, contributing paper for the sustainable infrastructure imperative: financing for better growth and development. New Climate Economy, London/Washington, DC
Godfrey L, Strydom W, Phukubye R (2016) Integrating the informal sector into the South African waste and recycling economy in the context of extended producer responsibility – CSIR briefing note. CSIR, Pretoria
Home R (2015) 4-colonial urban planning in Anglophone Africa. In: Urban planning in sub-Saharan Africa: colonial and post-colonial planning cultures, 1st edn. Routledge, New York
Hoornweg D, Bhada-Tata P (2012) What a Waste: A Global Review of Solid Waste Management. World Bank, Washington DC
Hosansky D (2018) Materials recovery facility. Encyclopædia Britannica, Inc.
Idowu IA, Atherton W, Hashim K et al (2019) An analyses of the status of landfill classification systems in developing countries: sub Saharan Africa landfill experiences. Waste Manag 87:761–771. https://doi.org/10.1016/j.wasman.2019.03.011
Kaza S, Yao L, Bhada-Tata P, Van Woerden F (2018) What a waste 2.0: a global snapshot of solid waste management to 2050. The World Bank, Washington DC
Knight Frank (2017) Real estate markets in a continent of growth and opportunity: Africa report 2017/18. Knight Frank LLP, London
Laffta S, Al-rawi A (2018) Green technologies in sustainable urban planning. MATEC Web Conf 162:05029. https://doi.org/10.1051/matecconf/201816205029
Lubida A, Veysipanah M, Pilesjo P, Mansourian A (2019) Land-use planning for sustainable urban development in Africa: a spatial and multi-objective optimization approach. Geodesy Cartogr 45:1–15. https://doi.org/10.3846/gac.2019.6691
Lwasa S, Kinuthia-Njenga C (2012) Reappraising Urban Planning and Urban Sustainability in East Africa. In: Polyzos S (ed) Urban Development. InTech
Mathye E (2019) Changing to a better, greener future. Department of Environmental Affairs, Pretoria

Mimura N, Pulwarty RS, Duc DM et al (2014) Chapter 15 – Adaptation planning and implementation. In: Climate change 2014: impacts, adaptation, and vulnerability. Part A: global and sectoral aspects. Contribution of working group II to the fifth assessment report of the intergovernmental panel on climate change. Cambridge University Press, Cambridge/New York, pp 869–898

Mohee R, Simelane T (eds) (2015) Future directions of municipal solid waste management in Africa. Africa Institute of South Africa, Pretoria

Nahman A, Godfrey L (2010) Economic instruments for solid waste management in South Africa: opportunities and constraints. Resour Conserv Recycl 54:521–531. https://doi.org/10.1016/j.resconrec.2009.10.009

Nzeadibe TC (2013) Informal waste management in Africa: perspectives and lessons from Nigerian garbage geographies: informal waste management in Africa. Geogr Compass 7:729–744. https://doi.org/10.1111/gec3.12072

Oduro-Appiah K, Afful A, Kotey VN, De Vries N (2019) Working with the informal service chain as a locally appropriate strategy for sustainable modernization of municipal solid waste management systems in lower-middle income cities: lessons from Accra, Ghana. Resources 8:12. https://doi.org/10.3390/resources8010012

Okeke DC, Nwachukwu MU (2019) Resilient Cities for New Regionalism in Africa. https://doi.org/10.5281/ZENODO.3459178

Oteng-Ababio M, Melara Arguello JE, Gabbay O (2013) Solid waste management in African cities: sorting the facts from the fads in Accra, Ghana. Habitat Int 39:96–104. https://doi.org/10.1016/j.habitatint.2012.10.010

Ozoike-Dennis P, Spaling H, Sinclair AJ, Walker HM (2019) SEA, urban plans and solid waste management in Kenya: participation and learning for sustainable cities. J Environ Assess Pol Manag 1950018. https://doi.org/10.1142/S1464333219500182

Reda AG, Tripathi NK (2011) Mainstreaming climate change adaptation in urban planning in Africa, p 12. Asia Institute of Technology, Chiang Mai

Revi A, Satterthwaite DE, Aragón-Durand F et al (2014) Chapter 8 – Urban areas. In: Climate change 2014: impacts, adaptation, and vulnerability. Part a: global and Sectoral aspects. Contribution of working group II to the fifth assessment report of the intergovernmental panel on climate change. Cambridge University Press, Cambridge/New York, pp 535–612

Sango T, Kissoon S, Basson L, Munganga G (2014) Development of the waste economy in the Western Cape: a decision support tool for integrated municipal waste management. Institute of Waste Management of Southern Africa, Somerset West, Cape Town

The Economist (2018) African cities pledge zero carbon emissions. https://country.eiu.com/article.aspx?articleid=536845837&Country=Kenya&topic=Economy. Accessed 23 Mar 2020

The World Bank (2017) Concept project information document integrated safeguards data sheet Greater Accra climate resilient and integrated development project P164330. The World Bank, Accra

Trois C (2014) Current RDI to find alternative solutions – anaerobic digestion of organic waste. Department of Science and Technology, Durban

Tuani E (2008) Managing urbanization and waste disposal in West Africa – feature article. The International Solid Waste Association, Accra

UNEP (2018) Africa waste management outlook. United Nations Environment Programme, Nairobi

Vaccari M, Tudor T, Vinti G (2019) Characteristics of leachate from landfills and dumpsites in Asia, Africa and Latin America: an overview. Waste Management 95:416–431. https://doi.org/10.1016/j.wasman.2019.06.032

Worldometer (2020) Population of Africa (2020). https://www.worldometers.info/world-population/africa-population/. Accessed 30 Apr 2020

Zafar S (2019) Introduction to materials recovery facility | EcoMENA. https://www.ecomena.org/materials-recovery-facility/. Accessed 23 Mar 2020

Zhang C, Xu T, Feng H, Chen S (2019) Greenhouse gas emissions from landfills: a review and bibliometric analysis. Sustainability 11:2282. https://doi.org/10.3390/su11082282

5

# African Yam Bean the Choice for Climate Change Resilience: Need for Conservation and Policy

C. V. Nnamani, D. B. Adewale, H. O. Oselebe and C. J. Atkinson

## Contents

Introduction .
Population Increase and Challenges of Food Security
Proffering Solution with Neglected and Underutilized Plant Genetic Resources
African Yam Bean an Underutilized Plant Genetic Resources
Nutritional Potentials of African Yam Bean for Climate Change Adaptation
Nomenclatural Etymology of African Yam Bean to Climate Change
African Yam Bean, the Choice for Climate Change Resilience .
Policy and Conservation African Yam Bean
Conclusions
References

C. V. Nnamani (✉)
Plant Systematics and Conservation Biology Research Unit, Department of Applied Biology, Faculty of Science, Ebonyi State University, Abakaliki, Nigeria
e-mail: nnamanicatherine@ebsu.edu.ng

D. B. Adewale
Department of Crop Science and Horticulture, Federal University Oye-Ekiti, Ikole-Ekiti, Nigeria

H. O. Oselebe
Department of Crop Production and Landscape Management, Ebonyi State University, Abakaliki, Nigeria

C. J. Atkinson
Natural Resources Institute, University of Greenwich, London, UK

Department of Agriculture, Health and Environment, Natural Resources Institute, University of Greenwich, Chatham, UK
e-mail: c.j.atkinson@gre.ac.uk; c.j.atkinson@greenwich.ac.uk

**Abstract**

Global warming has emerged as a major challenge to development and human wellbeing in Sub-Saharan Africa in general and Nigeria in particular. Periodic incidents show that this challenge will continue and increase in impact on all aspects of natural resources – agriculture, ecosystems services, biodiversity depletion, environmental degradation and human health. Recognizing the enormous potential of underutilized plant genetic resources (PGRs) is crucial as sources of solutions to a number of these threatening challenges emanating from climate change (food and nutrition insecurity, genetic erosion, loss of agro-biodiversity, green job growth and income generation) cannot be overemphasized. *Sphenostylis stenocarpa* (Hochst. *ex*. A. Rich) Harms., commonly known as African yam bean (AYB) belonging to the leguminous Fabaceae, is an underutilized PGR with rich portfolio which could serve as vital source of robust adaption and resilient germplasm for vulnerable local communities in Nigeria. Its substantial nutritional, environmental, cultural, social, medicinal, industrial and soil restorative potentials underpins its position as climate – smart species. Enhancing the potentials of African yam bean via robust innovative approaches for wider utilization through accelerated research, farmer seed exchanges, in-situ and ex-situ conservations, farmers selection, and policy programs such as seed sovereignty will accentuate its adaptation and used as resilient climate –smart species for the vulnerable groups in Nigeria to cushion impact of climate change.

**Keywords**

African yam bean · Climate smart species · Adaptation climate change · Resilience · Conservation · Nigeria

## Introduction

The Intergovernmental Panel on Climate Change (IPCC) reported that global warming of 1.5 °C and greater upsurge of 2 °C is projected for 2046–2065 and that these will have consequential impacts on food and water security, health, and other components of sustainable development, limiting strategies for adaptation, losses and damages to biodiversity and ecosystem services (IPCC 2018). All players are potential culprits to these increase, and accordingly need to undertake considerable behavioral changes to counteract this life threatening global warming, particularly in Sub-Saharan Africa. However, the above scenario is dependent on the level of anthropogenic activities on future greenhouse gas emissions (Pachauri and Meyer 2014).

Resilience is required within any systems to deal with climate stresses and disturbances and this include the willingness to change, skills to learn and capacity to adapt to such changes. It is therefore the zeal of managing changes and adaptations that can stand the test of current and future climate risks.

## Population Increase and Challenges of Food Security

There has been a geometric increase in global population over the last 200 years and it is projected to continue to rise. United Nations forecast stated that due to exponential growth, that population will rise from the current 7 billion to reach 9.7 billion by 2050 and subsequently 10.9 billion by 2100 (Fig. 1). This brings with it a number of life threatening challenges, especially in the absence of adequate food production technology, integrated programs which simultaneously address those community needs for food and reproductive health and impact of climate variability and change (World Bank 2014). The consequential effects of this population increase will be greater in Sub-Saharan Africa due to many factors such as, higher fertility rate, demographically younger population entering childbearing life phase, joblessness and a more volatile employment market.

Rapid national population growth has induced and intensified major environmental changes in response to climate change and variability which have led to the emergence of novel pests and diseases, loss of crop genetic diversity and a decline in resources such as water (Dobermann and Nelson 2013). Despite these, current global food production rates being at there highest level, they still fail to match population demand. Given the predicted population increase, food production will need to increase to meet demand by between 59% and 98% by 2050 (Elferink and Schierhorn 2016). United Nations Food and Agriculture Organization (FAO 2010) estimated that about 795

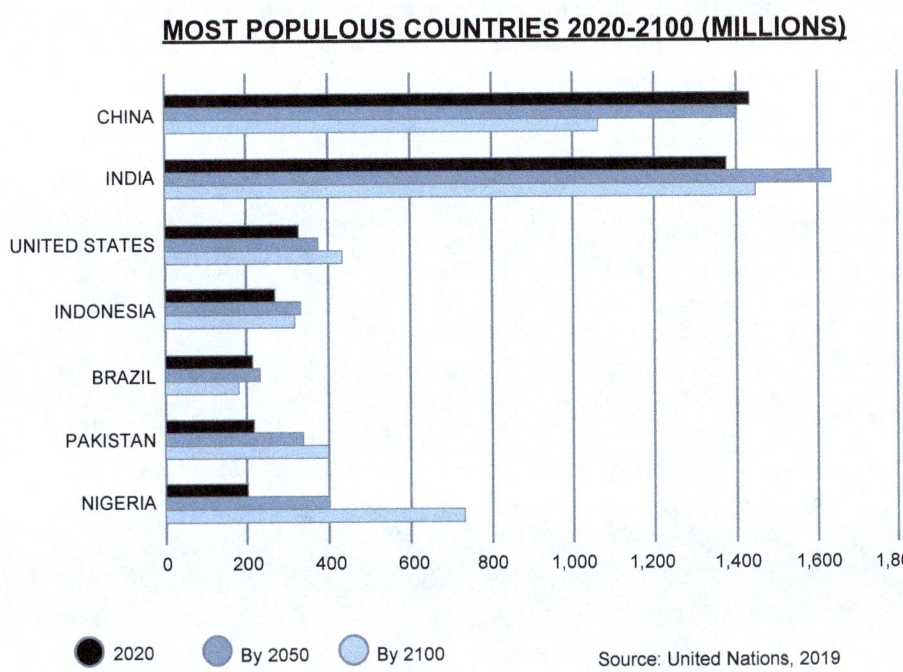


**Fig. 1** Predicted population growth for 2020–2100

million people, of the current 7.3 billion population, suffered chronic malnourishment between 2014 and 2018 and majority of these people lived in developing countries. The Global Hunger Index (Grebmer et al. 2019) shows that multiple countries have higher hunger levels now than in 2010, and approximately 45 countries in current reality will not be able to achieve lower levels of hunger by 2030.

Poor monotonous diets, low in calorific quantity and quality with respect to nutritional level, variety and diversity are the primary cause for poor human health in many developing countries. Coincidentally, while the number of mouths to feed is increasing more rapidly than those farmers who are able to produce food. Hunger, malnutrition and famines and many endemic diseases, desertification, and floods continue to pose high risks in developing countries with low resilience which will be further exacerbated by climate change and climate variability aboard (Dobermann and Nelson 2013).

Nigeria ranks 93rd out of 117 qualifying countries with a score of 27.9, showing that its population is malnourished with high number of hungry people. This is an indication that out of the 162.5 million people, 1 in every 7 go to bed hungry (Grebmer et al. 2019). Mary Robinson, an Adjunct Professor of Climate Justice, Trinity College Dublin and former UN High Commissioner for Human Rights stated that "it is a terrible global indictment that after decades of sustained progress in reducing global hunger, climate change and conflict are now undermining food security in the world's most vulnerable regions." She concluded that with the number of hungry people rising from 785 million in 2015 to 822 million in 2018, "we can no longer afford to regard the 2030 Agenda and the Paris Climate Agreement as voluntary. This could be inferred from the level of changes over time in West Africa (Fig. 2). The above scenario needs vital actions and collective change of

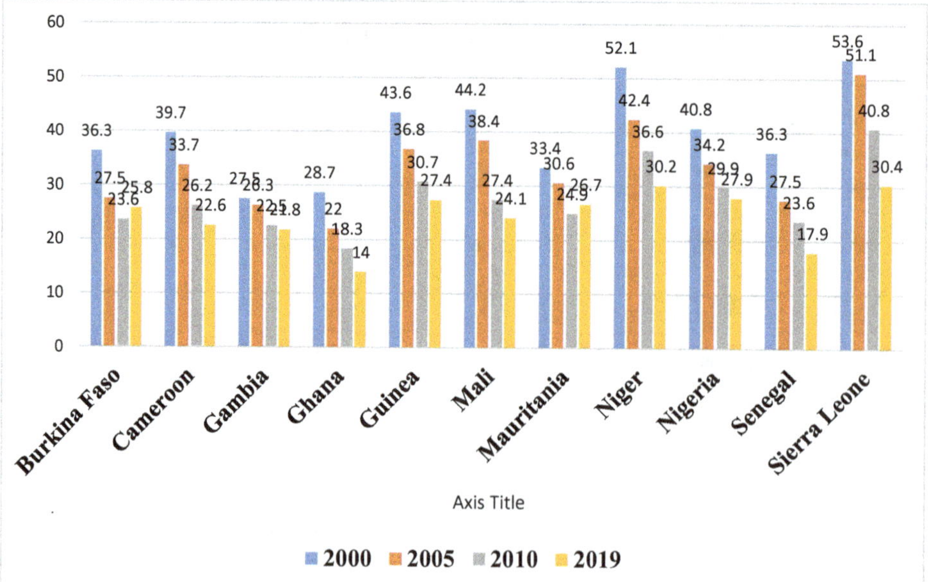


**Fig. 2** West Africa Global Hunger Index scores, and changes over time. (Source: Grebmer et al. 2019)

attitude at both global and local levels in order to secure a habitable world for our forthcoming generations" (Grebmer et al. 2019).

## Proffering Solution with Neglected and Underutilized Plant Genetic Resources

Plant genetic resources (PGR) are those varieties of heritable materials confined in traditional selections and modern domesticated cultivars together with their wild relatives and other wild plant species locally used as food, feed for domestic animals, fibre, clothing, medicine, shelter, wood, timber, energy and for recreational activities.

Neglected and underutilized plant genetic resources are domesticated and wild plants species which have been used for centuries as sources of food, fibre, fodder, oil, or have medicinal properties (stimulants, narcotics, aromatics), but their importance has remained unharnessed, abandoned, replaced with exotic foods or only very slowly now being considered for use. They are known by various names, including orphan, abandoned, lost, forgotten, neglected, underutilized, local, minor, poor man's food, traditional, alternative, niche, or underdeveloped plants (Padulosi 2017). Their underutilized potentials can be significantly exploited to combat climate change, as a future "smart crop" to improve food and nutrient insecurity. Here we adopt a progressive and positive approach to their potential exploitation as a source of "smart food," with an emphasis on their exploitation and/or development into viable future crops which are climate resilient. The Global Facilitation Unit for Underutilized Species (Bhat and Taiwo 2015) defines these species, as those with a potential not fully exploited; with respect to contributing to food security and poverty alleviation, combined with their strong links to local community cultural heritage. Their limited or poor documentation by research are evidenced by their adaptation to specific agro-ecological niches, a weak or non-existent seed supply systems, fully in the domains of traditional users only. They are produced with little, or no external inputs.

Our ancestors used over 7000 plant genetic resources (PGRs) as food, from which farmers produced about 70% of the global food supply (Bhat and Taiwo 2015). Currently, very few of these PGRs are still exploited in sustaining humans. Making food and nutrient security dependent on a handful of crops with 65% requirements for protein, calories and minerals met by wheat, rice, yam and cassava (FAO 2017). This shrinking portfolio of PGRs, shifts in life style, and the preference for fast foods are currently contributing factors involved in food and nutrient insecurity. These are reducing the capacity of farmers to adapt to changing markets, to innovate, to develop supply chains, and our ability to cope with climate change variability and its extremities (IPGRI 2003). Recognizing the considerable potentials which underutilized PGR can provide is crucial in finding solutions to the impacts of climate change on food and nutritional insecurity, genetic erosion, conservation of agro-biodiversity, green growth jobs and income generation (Padulosi et al. 2013; Nnamani 2015).

Nigeria as a nation holds a rich variety of PGRs linked to food, and other treasured traits. However, most of these species have received little, or no attention with respect to support for basic research, breeding improvement, commercial exploitation of germplasm selections, or consideration by policy itineraries to support enhanced utilization, thereby leaving them neglected. In order to meet food security challenges for a growing population under changing climate, it is necessary to identify and valorize these neglected and underutilized species (NUS) to develop greater resilience within local communities and the adaptation potential to increasing climate unpredictability and change.

## African Yam Bean an Underutilized Plant Genetic Resources

*Sphenostylis stenocarpa* (Hochst. ex. A. Rich) Harms, customarily called African yam bean (AYB), is a legume of the subfamily Faboideae and family Fabaceae. *Sphenostylis* as a genus is represented by only 7 species (Potter and Doyle 1992). African yam bean is an orphan crop cultivated and utilized by local communities. It is a climbing annual, prostrate, or erect and approximately 1–3 m tall. The leaves are trifoliate, about 2.7–13 cm long and 0.2–5.5 cm broad. The inflorescence could be a raceme which shows an acropetal variety of flowering with pink colour intermingled with purple, twisted backward characteristics of the Fabaceae family (Fig. 3).

African Yam Bean is synonymous with African continent only, its geographical distribution is tolerated mainly within the soil, climate, and vegetation of Africa especially within the range of latitudes 15° North to 15° South and longitudes 15° West to 40° East (Adewale et al. 2008). The above geographical range could be referred to as the center of diversity of AYB (Fig. 3) since there is no record of the origin of the crop in any other continent beyond Africa (Potter and Doyle 1992). It is a crop of African origin, made for Africans and utilized as such in Africa, hence this

**Fig. 3** Growth habit, floral and fruits features of African yam bean.

confirming the common claim that the crop is a tropical African legume. The rich hot spots of *S. stenocarpa* is found in Chad and Ethiopia in northeast tropical Africa, Kenya, Tanzania and Uganda for east tropical Africa, and in Burundi, Central African Republic and Zaire for west-central tropical Africa. It is also found in West Africa in places like Cote d'Ivoire, Ghana, Guinea, Mali, Niger, Nigeria and Togo while in south tropical Africa it is cultivated in Angola, Malawi, Zambia and Zimbabwe (USDA Agricultural Research Service 2015) (Fig. 4).

In South-east, Nigeria, the plant is diverse in distribution and geographical range thriving more in the derived savanna vegetation zone (Fig. 5).

## Nutritional Potentials of African Yam Bean for Climate Change Adaptation

The seeds and tubers of AYB are highly rich in protein, minerals, and vitamin (Fig. 5). It is a promising source of plant protein for resource poor rural and semi-rural communities. The amino acid (lysine and methionine) has been reported to be higher than those of pigeon pea, cowpea, and bambara groundnut (Uguru and Madukaife 2001). Omeire (2012), noted that the amino acid (g $100^{-1}$ g) profile of AYB comprised of lysine (6.1), histidine (3.1), arginine (6.5), aspartic acid (9.1), glycine (3.9), alanine (4.1), valine (5.0), and phenylalanine (5.1) (Fig. 5). Its lysipne and methionine contents were similar or better than those of soybean protein (Yetunde et al. 2009) and comparable with whole chicken eggs (Ekpo 2006). The protein profile of the tubers compares favourably with that of other African root crops such as yams and sweet potatoes and has almost two and half times the protein value of cassava tubers (National Research Council 1979; Amoatey et al. 2000; Adewale and Aremu 2013) (Fig. 6).

African yam bean is cultivated for its edible seeds and tubers (Fig. 7), which have sustained indigenous community's livelihoods. A report of its medicinal value was reported in a traditional Igbo settings at Enugu State, Nigeria. The seed is an important ingredient utilized in the topical treatment of stroke, insomnia, diabetics, measles and stress. Extract of mashed AYB, after cooking, is used to induce lactation in mothers after child birth, while the fried seed coat is ground and used in the treatment of stroke (Nnamani et al. 2017). Culturally, it is used traditionally to entertain guests during traditional marriage ceremonies and serves as a famine food when every other crops have failed. Industrially, African yam bean flour is used to fortify conventional wheat flour which is frequently lower in protein content (Nnamani et al. 2017).

## Nomenclatural Etymology of African Yam Bean to Climate Change

*Sphenostylis stenocarpa* is botanically synonyms in numerous literatures with *Sphenostylis ornata* A. Chev., *Sphenostylis congensis* A. Chev., *Dolichos stenocarpus* Hochst. ex. A. Rich., *Sphenostylis katangensis* (De Wild.) Harms, *Vigna katangensis* De Wild., and *Vigna ornata* Welw. ex. Baker. However, the

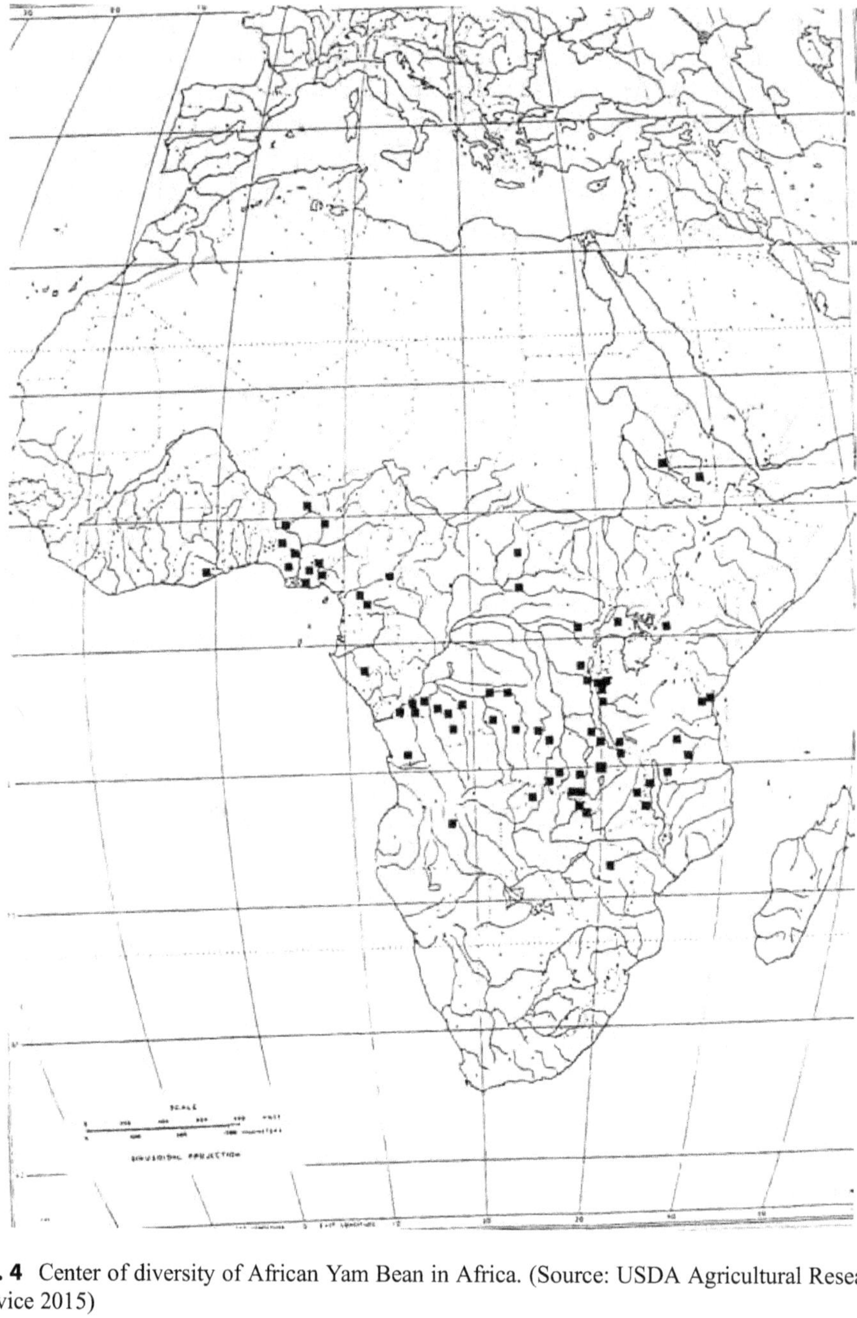

**Fig. 4** Center of diversity of African Yam Bean in Africa. (Source: USDA Agricultural Research Service 2015)

current acceptable nomenclature is *Sphenostylis stenocarpa* (Hochst. ex. A. Rich) Harms (Adewale et al. 2012). Keay (1989) reported the followings common names as "Diegemtenguere" (Mali), "Girigiri" (Hausa, West Africa), "Norouko" and/or "Roya" (Sudan). Adewale and Odoh (2013) reported on some of the tribal synonyms of the crop in Nigeria: it is known as "Azama" (Ebonyi, State), "Akidi" or "Azima" or "Okpududu" (Ohafia, Abia State) and "Uzaaku" or "Ijiriji" (Nsukka, Enugu State)

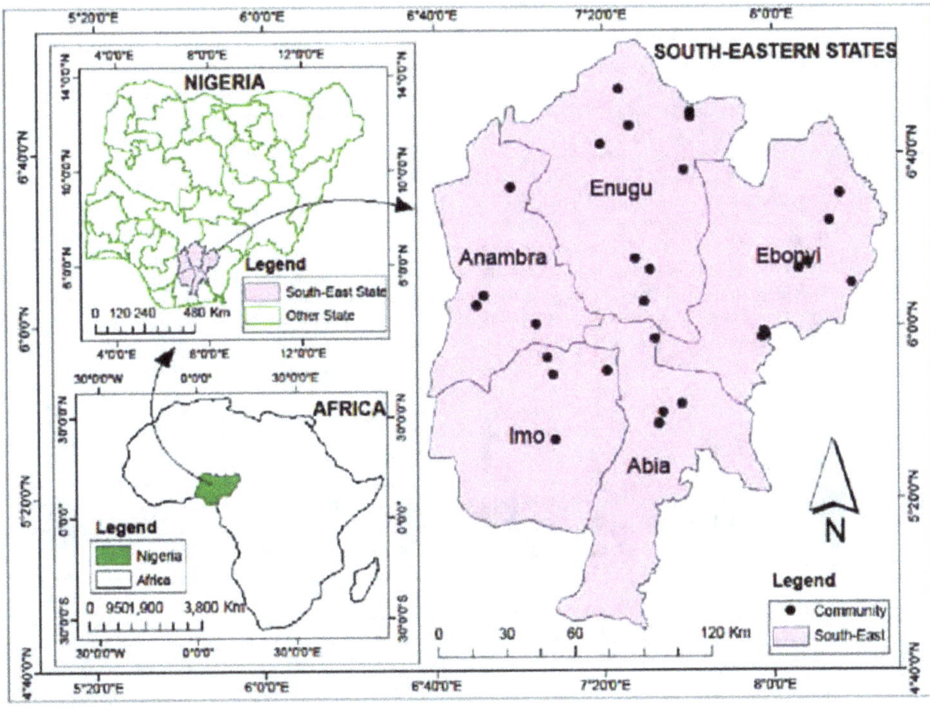


**Fig. 5** Geographical distribution of African yam bean in south-east Nigeria. (Source: Nnamani et al. 2017)

in Igbo land; it is "Ewe" (Ijesha, Osun State), "Otiili" (Ekiti, Ekiti State),"Ekulu" (Ipe-Akoko, Ondo State), "Orogodo" (Owo, Ondo State), "Peu" (Ijebu, Ogun State) and "Sunmunu" (Iseyin, Oke-Ogun, Oyo State), Yoruba land. Some other tribal names for AYB in Nigeria are "Ihiehie" (Ishan, Edo State), "Iye" (Estako, LGA, Edo State),"Ahuma" (Tiv, Benue State), "Nsama" (Efik-Ibibio, Akwa Ibom) and Cross River State) (Adewale and Odoh 2013).

Al Azharia (2006) noted that throughout human history, folk nomenclature has been the underpinning factor for plant selection, improvement and conservation. It was and still is customarily for the local custodians to recognized species which have accomplished high serviceable and cultural values by custodians of such plants worldwide. When folk names given to species it often carries etymological meaning such as healing effects, morphological features, mythical connotations, adaptive strength, mitigating capability and allegorical values (Nnamani et al. 2019). Etymological data showed the AYB had sustained the lives of the local communities for ages. Indigenous lingual synonyms for AYB in Nigeria have etymological significances relating to how this crop had interfaced with the livelihood and environmental conditions of the local communities in different scenarios in the past (Table 1). These names often convey relationships to specific plant therapeutic effects, nutritional capacity, morphological features, mythological connotations, and ecological conditions playing out at a particular point in time in their environs (Leyew 2011; Rankoana 2012).

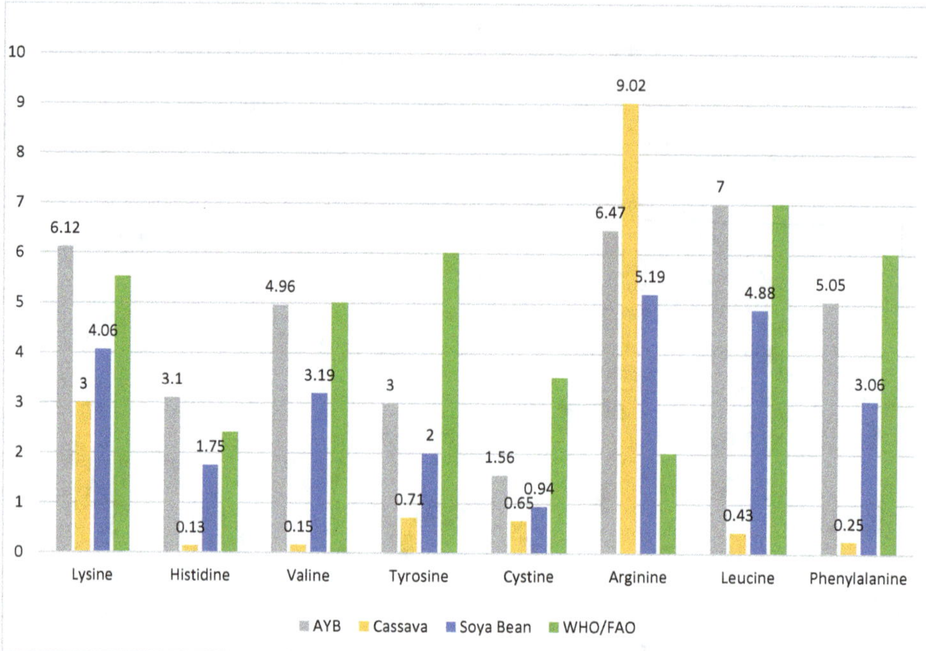


**Fig. 6** Amino acid profile of African yam bean compared with other crops and WHO/FAO bench marks. (Source: Omeire 2012)

**Fig. 7** Seeds and tuber of African yam bean.

## African Yam Bean, the Choice for Climate Change Resilience

"Neglected and Underutilized species (NUS)" are plant species which are sidelined, ignored or completely underexploited, by researchers, breeders and policy makers. They have great prospects to improve people's livelihoods, guaranteeing food security and self-sufficiency to the users. Their environmental restoration ability are not yet fully appreciated due to their limited competitiveness with fundamental

**Table 1** Etymology of African yam bean in South-east, Nigerian

| S/ no. | Ethnic group | Name | Meaning | Etymological facts |
|---|---|---|---|---|
| 1 | Izzi | Azama | Feeder of people in hard times (mitigating food) | They believed that God had answered their prayer by giving them this crop to feed them when other crops have failed in the land |
| 2 | Ezza | Azama | Healing food for the people (medicine) | In the past when kwashiorkor (malnutrition/hidden hunger was the major problem facing the people of Ezza, boiling the crop, eating AYB and the use of water from the cooked seeds to bath provides healing |
| 3 | Ikwo | Azaaki | Multiple seed-food provider (resilient seed) | They named AYB Azaaki because of the multiple food types it provides, which help poor farmers when they cultivate the crop. They believe that cultivating AYB is a way of avoiding hunger |
| 4 | Ngbo | Azaakuru | Plant small and harvest plenty (dominating and climate–smart crop) | Aza-akuru because, it is a passive crop but when planted, it will dominate the whole farm and yield lots of food types both seed and the tuberous root. A dominating crop that God gave to help poor farmers during famine |
| 5 | Ntezi and okpoto | Etiti | "Subduer of hunger" (the central hungry killer) | They believed that God has given them the crop to feed them during a time of famine. AYB is the only crop that can feed people even when the quantity is small |
| 6 | Afikpo | Azuma | Food server<br>God and medicine | They name this crop Azuma because of its great feeding capacity and its ability to stop some illness, e.g., stomach disorder. They believed that God answered their prayer in times past |
| 7 | Ohozara | Aza-ama | Dominating food crop<br>God and medicine | Prayer answered by God for the crop provides food even when other crops have failed. It yields great food for the poor. When the poor cannot cultivate other crops as a result of environmental changes |

(continued)

**Table 1** (continued)

| S/ no. | Ethnic group | Name | Meaning | Etymological facts |
|---|---|---|---|---|
| 8 | Nkalagu | Azaku | Food helper of the poor | They called it "Azaku" because of the role it played during famine in times past. They believed that God provided it to help the poor through its simplicity in cultivation and survival |
| 9 | Ohafia-Agba | Azama | Feeder of people<br>Adaptation to climate change | They believed that the God of harvest has provided this crop to sustain his people when other crops do not perform very well on the farm. AYB can compete in and adapt to many difficult environments |
| 10 | Agba, Nkalegu and Igboasa | Eriwa | Minor crop that feeds many people<br>Gender | It's a women's crop that helps to sustain the women's little income when the major crops (the men's crops like yam, rice and cassava) are finished in the bans, particularly after planting. AYB is usually intercropped with other crops by women hence it is termed women's crop |
| 11 | Ezza and Ikwo | Onyiaoduru | Sustainer and friend of the farmer<br>Sustainer | Respondents believe that after eating AYB, the consumer keeps drinking water through the day and it gives sufficient energy to work without getting exhausted |
| 12 | Ngbo and Izzi | Mgbadumue | Saving seed<br>Seed | When labourers are hired to work on farms. They eat the food in the morning and it will keep them active until evening |

Source: Nnamani et al. (2019)

crops in conventional use. Although their potential may not be fully realized at national or international levels, they are of significant importance locally, because of their high adaptation capacity to marginal, complex, and difficult environments contributing significantly to diversification and resilience of agroecosystems (Padulosi et al. 2011).

Shyam et al. (2014) stated that climate modeling is the commonest approach used today for predicting the responses of species to climate change. Species which have resilience biotic characteristics are often informative when used in climate modelling and underutilized plants frequently have an array of stress resistance traits. They are of significant importance locally, highly adaptable to marginal soil, complex and changing environments and agriculturally degraded land. They are hubs contributing meaningfully to diversification and resilience to community agro ecosystems, this is because of their long-standing use with little, or no, conventional breeding improvement. As orphan crops, notwithstanding their intrinsic low-yield potential, they still have the capacity of resisting climate stresses, having been selected and exploited in local climate stress adaptation within the local communities for generations. In this regard, underutilized PGR could contribute to further building of community resilience over the long-term in response to impending climate change adaptation after effective valorization of a PGR's potential (Mabhaudhi et al. 2019).

African yam bean usefulness as a climate smart species for communities to achieve resilience to climate stresses is based on its ability to survive and even thrive (reliable yield annularly) in marginal soils and environments where other crops fail in the absence of often unavailable resources, such as nitrogen fertilization and adequate irrigation. AYB, as a legume, improves soil quality and function through its beneficial effects on soil biological, chemical and physical attributes. Such benefits include enabling improvements in crop N-supply, increasing soil organic matter reserves, stimulating soil microbial diversity and activity, and improvements in soil structure, reducing soil erosion while increasing soil aeration and enhanced soil water-holding capacity. AYB is suitable for wide climatic conditions (Anochili 1984; Betsche et al. 2005) and nitrogen fixation (Oganale 2009; Tettey 2014). Ojuederie and Balogun (2019) obtained better tuber production from genotypes in a year with lower rainfall and sunshine hours. This is a display of smart features.

## Policy and Conservation African Yam Bean

Intergovernmental Panel on Climate Change (IPCC 2018) reported the likely consequences of rapid global climatic change on species and their habitats. Such changes could cause a shift in the conservation status of so many species, pushing some towards the red line list of International Union for Conservation of Nature (IUCN). African yam bean has not received much conservation attention to date. Currently, much of its germplasm is in the hands of the local custodians as landraces (Nnamani et al. 2018), while the Genetic Resources Centre at the International Institute of Tropical Agriculture (IITA), Ibadan, Nigeria also hold some quantity of AYB germplasm. The disadvantages of poor conservation technology include the limited access to germplasm by researchers, breeders and farmers, the vulnerability of AYB to natural disasters, genetic erosion and loss of crop diversity are there.

Enhancing the potentials of African yam bean by policy via robust innovative approaches for wider utilization through accelerated research, farmer seed exchanges, in-situ and ex-situ conservations, farmers selection, and policy programs such as seed sovereignty will accentuate its adaptation and used as resilient climate – smart species for the vulnerable groups in Nigeria to cushion the impact of climate change.

These species may be conserved in-situ or ex-situ for future generations. On-farm (in-situ) techniques have several advantages compared with ex situ conservation, as the former allows for evolution of traits through continued natural and human-driven selections, which will contribute to greater crop adaptation and resilience to abiotic climate stress (Padulosi 2017).

Through the recognition of its promising nutritional potentials, policy could be synchronized in the conservation, promotion, marketing network and utilization of AYB by mainstreaming it into various dietary menus and products to include a range of food products appropriate for a broad range of consumption opportunities. This should include a rebranding of its nomenclature from "poor man's food" to a more widely acceptable range of producers, retailers and consumers. Decision makers could equally support the development of value chains, scale up development collaborations and information sharing of AYB among researchers, conservationists, breeders, extension workers, farmers, farmer organizations, vendors and NGOs.

Finally, there is need for policy to intensify efforts to support research on the conservation of *S. stenocarpa* through participatory engagement between the species custodians of the species for both in situ and ex situ conservation approaches. Globally the exploitation and development of legumes, which are climate resilient, will provide a significant opportunity for improving human health through integration into food-systems as a replacement, or partial replacement, of current animal sourced proteins. A programmed germplasm conservation approach is needed to collect, characterize, develop and exploit existing diversity in the provision of food, feed, nutritional security and environmental resilience.

## Conclusions

Convincingly, recognizing the potential of AYB as an excellent climate smart species through active food governance policy would initiate novel business opportunities by changing the narrative via:

(i) Scaling up programs on the use of African yam bean in its various dietary forms and consumptions routes. This could rebrand the nomenclature of the species from "poor man's food" to an added value nomenclature (high protein source).
(ii) Organizing regular retailer food fairs to showcase various dietary menus accruing from AYB will encourage its selection and consumption by the publics.
(iii) Support the development of value chains and market networks for small agribusinesses for priority crops such as AYB.
(iv) Provide credit facilities and farmer/user-friendly technologies to the crop custodians to accelerate production.
(v) Promote the use of AYB through directed awareness campaigns on the nutritional benefits of the crop.
(vi) Exploit the offsetting potential AYB consumption of added value of this nutritious crop against the probable reduction in government health budget associated with alleviating malnutrition and it related conditions.

(vii) Including policy programs that enhance the selection of quality AYB seed available to farmers, with respect to climate resilience.
(viii) AYB will require more attention firstly, research-based, then technological germplasm development and subsequently supporting exploitation policy and markets. This will direct and facilitate appropriate solutions to major constraints for greater AYB cultivation and utilization.
(ix) Scaling up development via policy programs has the prospect of empowering resource poor communities who are predominantly women and the unemployed.
(x) Promote direct and appropriate consumer lead engagement and communication between knowledge acquirers and exploiters to drive AYB germplasm conservation.
(xi) Innovative biotechnological tools should be used to catalyze transformative change in AYB to promote its selection as a major staple food through a broad food product base.

**Acknowledgements** This work was supported by the CIRCLE Visiting Fellowship Program funded by UK Department for International Development (DfID). Neither the findings none the views expressed, however, necessarily reflects the policies of UK Government.

## References

Adewale BD, Aremu CO (2013) The Nutritional Potentials and Possibilities in African yam bean for Africans. Int J Agric Sci 3(1):8–19

Adewale BD, Odoh NC (2013) A review on genetic resources, diversity and agronomy of African yam bean (*Sphenostylis stenocarpa* (Hochst. ex. A. Rich.) Harms): a potential future food crop sustainable. Agric Res 2:3–43

Adewale BD, Kehinde OB, Odu BO, Dumet DJ (2008) The potentials of African yam bean (*Sphenostylis stenocarpa* Hochst. ex. A. Rich) Harms in Nigeria: character distribution and genetic diversity. In: Smart J, Haq N (eds) New crops and uses: their role in a rapidly changing world. Centre for underutilized crops. RPM Print & Design, Chichester, pp 265–276

Adewale BD, Dumet DJ, Vroh-Bi I, Kehinde OB, Ojo DK, Adegbite AE, Franco J (2012) Morphological diversity analysis of African yam bean and prospects for utilization in germplasm conservation and breeding. Genet Res Crop Evolut 59:927–936

Al Azharia J (2006) How Plant Names Reveal Folk Botanical Classification, Trade, Traditional Uses and Routes of Dissemination (II), in Asian Studies. Int J for Asian Stud 7:77–128

Amoatey HM, Klu GYP, Bansa D, Kumaga FK, Aboagye LM, Benett SO, Gamedoagbao DK (2000) African yam bean (*Sphenostylis stenocarpa*): a neglected crop in Ghana. West Afr J App Ecol 1:53–60

Anochili BC (1984) Tropical agricultural handbook. In: Food crop production. Macmillan Publishers, London, pp 48–50

Betsche T, Azeke M, Buening-Pfaue H, Fretzdorff B (2005) Food safety and security: fermentation as a tool to improve the nutritional value of African yam bean. In: Conference proceeding of the international agricultural research for development, Stuttgart–Hohenheim, October 2005, pp 1–5

Bhat JA, Taiwo S (2015) Plant genetic resources and traditional knowledge in light of recent policy developments. In: Salgotra R, Gupta B (eds) Plant genetic resources and traditional knowledge for food security. Springer, Singapore

Dobermann A, Nelson R (2013) Opportunities and solutions for sustainable food production, background research paper. High-level panel of eminent persons on the post-2015 development agenda. www.post2015hlp.org/wp-content/uploads/2013/05/Doberman-Nelson_Solutions-for-Sustainable-Food-Production.pdf

Ekpo AS (2006) Changes in amino acid composition of African yam beans (*Sphenostylis stenocarpa*) and African locust beans (*Parkia filicoida*) on cooking. Pak J Nutr 5:254–256

Elferink M, Schierhorn F (2016) Global demand for food is rising. Can we meet it?. Harvard Business Review. https://www.researchgate.net/publication/302466629_Global_Demand_for_Food_Is_Rising_Can_We_Meet_It

FAO (2010) The State of Food Insecurity in the World Addressing food insecurity in protracted crises. Rome. http://www.fao.org/3/a-i1683e.pdf

Food and Agricultural Organization of the United Nations (FAO) (2017) The state of food insecurity in the world 2017. FAO, Rome

Germplasm Resources Information Network (GRIN) (2009) GRIN taxonomy for plants (online)

Grebmer K, Bernstein J, Mukerji K, Patterson F, Wiemers M, Ní Chéilleachair R, Foley C, Gitter S, Ekstrom K, Fritschel H (2019) Global Hunger Index: The Challenge of Hunger and Climate Change. Bonn: Welthungerhilfe; and Dublin: Concern Worldwide

International Plant Genetic Resources Institute (IPGRI) (2003) Annual report. International Plant Genetic Resources Institute (IPGRI), Rome. ISBN 92-9043-591-7

IPCC (2018) Summary for policymakers. In: Masson-Delmotte V, Zhai P, Pörtner HO, Roberts D, Skea J, Shukla PR, Pirani A, Moufouma-Okia W, Péan C, Pidcock R, Connors S, Matthews JBR, Chen Y, Zhou X, Gomis MI, Lonnoy E, Maycock T, Tignor M, Waterfield T (eds) Global warming of 1.5 °C. An IPCC special report on the impacts of global warming of 1.5 °C above pre-industrial levels and related global greenhouse gas emission pathways, in the context of strengthening the global response to the threat of climate change, sustainable development, and efforts to eradicate poverty. World Meteorological Organization, Geneva, p 32

Keay RWJ (1989) Trees of Nigeria. Clarendon Press/Oxford University Press, New York, p 476

Leyew Z (2011) Wild plant nomenclature and traditional botanical knowledge among three ethnolinguistic groups in northern Ethiopia, Ossrea edn. Ossrea, Addis Abba

Mabhaudhi T, Chimonyo V, Hlahla S (2019) Prospects of orphan crops in climate change. Planta 250:695–708. https://doi.org/10.1007/s00425-019-03129-y

National Research Council (1979) Tropical legumes: resources for the future. National Academy of Sciences, Washington, DC, p 246

Nnamani CV (2015) Innovative stride to zero hunger beyond 2015 in Nigeria. Recent Pat Food Nutr Agric 7:14–21

Nnamani CV, Ajayi SA, Oselebe HO, Atkinson CJ, Igboabuchi AN, Ezigbo EC (2017) *Sphenostylis stenocarpa* (ex. A. Rich.) Harms., a fading genetic resource in a changing climate: prerequisite for conservation and sustainability. Plants 6:30. 1–16. https://doi.org/10.3390/plants6030030

Nnamani CV, Ajayi SA, Oselebe HO, Atkinson CJ, Adewale DB, Igwe DO, Akinwale RO (2018) Updates on nutritional diversity in *Sphenostylis stenocarpa* (Hoechst ex. A. Rich.) Harms., for food security and conservation. Am J Agric Biol Sci 13:38–49. https://doi.org/10.3844/ajabssp.2018.38.49

Nnamani CV, Atkinson CJ, Ejike JE (2019) Etymology of folk nomenclatures for *Sphenostysis stenocarpa* (Hoechst ex. A. Rich.) Harms. J Underutil Legum 1(1):135–147

Oganale, D. (2009). Nodulation and nitrogen fixation of African yam beam (*Sphenostylis stenocarpa*). In: Proceedings of African Crop Science Society conference, Cape Town, 28 September–1 October 2009. Science and technology supporting food security in Africa

Ojuederie OB, Balogun MO (2019) African yam bean (*Sphenostylis stenocarpa*) tubers for nutritional security. J Underutil Legum 1(1):56–68

Omeire GC (2012) Amino acid profile of raw and extruded blends of African yam bean (*Sphenostylis stenocarpa*) and cassava flour. Am J Food Nutr 2:65–68

Pachauri RK, Meyer LA (2014) IPCC climate change 2014: synthesis report. Cambridge University Press, Cambridge

Padulosi S (2017) Bring NUS back to the table! GREAT insights Magazine 6(4):21–22. https://hdl.handle.net/10568/89038

Padulosi SN, Heywood V, Hunter D, Jarvis A (2011) Underutilized species and climate change: current status and outlook. In: Yadav SS, Redden RJ, Hatfield JL, Lotze-Campen H, Hall AE (eds) Crop adaptation to climate change. Wiley-Blackwell, Oxford, pp 507–521. https://doi.org/10.1002/9780470960929.ch35

Padulosi S, Bergamini N, Lawrence T (2013) On farm conservation of neglected and underutilized species: status, trends and novel approaches to cope with climate change. In: Proceedings of an international conference of biodiversity international, Frankfurt, 14–16 June, 2011. Biodversity International, Rome

Potter D, Doyle JJ (1992) Origin of African yam bean (*Sphenostylis stenocarpa*, Leguminosae): evidence from morphology, isozymes, chloroplast DNA and linguistics. Econ Bot 46:276–292

Rankoana SA (2012) The use of indigenous knowledge for primary health care among the Northern Sotho in the Limpopo Province. Unpublished Ph.D. thesis, University of Limpopo, Sovenga

Shyam SS, Zachariah PU, Kripa V, Mohan A, Ambrose TV, Manju RS (2014) Vulnerability assessment of coastal fisher households in Kerala: a climate change perspective. Indian J Fish 61:99–104

Tettey A (2014) Nodulation, nitrogen fixation and diversity of the African yam bean rhizobia in four Ghanaian soils. M.Phil. thesis. University of Ghana. http://197.255.68.203/handle/123456789/7097

Uguru MI, Madukaife SO (2001) Studies on the variability in agronomic and nutritive characteristics of African yam bean (*Sphenostylis stenocarpa* Hochst. ex. A. Rich. Harms). Plant Prod Res J 6:10–19

USDA Agricultural Research Service (2015) Germplasm Resources Information Network (GRIN). Ag Data Commons. https://doi.org/10.15482/USDA.ADC/1212393. Accessed 2020-11-29.

World Bank (2014) Turn down the heat: confronting the new climate normal. World Bank, Washington, DC. License: creative commons attribution – non commercial – no derivatives 3.0 IGO (CC BY-NC-ND 3.0 IGO).http://documents.worldbank.org/curated/en/317301468242098870/pdf/927040v20WP00O0ull0Report000English.pdf

Yetunde EA, Ukpong SU, Olajumoke L, Ime FA (2009) Nutrient composition and sensory properties of cakes made from wheat and African yam bean flour blends. J Food Technol 7:115–118

6

# Smallholders use of Weather Information as Smart Adaptation Strategy in the Savannah Area of Ondo State, Nigeria

Rasheedat Alliagbor, David Olufemi Awolala and Igbekele Amos Ajibefun

## Contents

Introduction ....
Materials and Methods
Econometric Model
Heckman Two-Step Procedure ...
Results and Discussions
Drivers of Farmers' Use of Weather Information for Smart Decisions
Conclusions and Policy Recommendations
References

**Abstract**

Weather information is needed for smart decisions because uncertainties in weather phenomena are beyond capacity of smallholders for autonomous responses. The study analyzed determinants of farmers' use of weather information as smart adaptation strategy. Arable crop farmers were sampled in the derived

R. Alliagbor (✉)
Department of Agricultural and Resource Economics, Federal University of Technology Akure, Akure, Ondo, Nigeria
e-mail: jessicaalliagbor@gmail.com

D. O. Awolala
African Climate Change Leadership Program (AfriCLP), University of Nairobi, Kenya and Institute of Resource Assessment, University of Dar es salaam, Dar es Salaam, Tanzania
e-mail: ddawolala@gmail.com

I. A. Ajibefun
Department of Agricultural Economics and Rural Sociology, Auburn University, Auburn, AL, USA
e-mail: iajibefun@yahoo.com

savannah agro-ecology area of Ondo State, Nigeria. Heckman probit model estimates show that gender and agricultural extension services were significant positive drivers of farmers' access to weather information. Should weather information becomes an alternative adaptation strategy, access to credit was found as the major driver of farmers' propensity to use weather information before taking climate smart agricultural decisions. Further results reveal that increasing knowledge of onset date, large farm size, and access to agricultural extension services significantly reduced farmers' propensity to use improved weather information for smart decisions in the dry savannah area.

**Keywords**

Extreme weather · Weather information · Adaptation responses · Smart decisions · Heckman Probit · Ondo State

## Introduction

Climate change is a particular threat to continued economic growth and livelihoods of vulnerable populations (UNEP 2018). The United Nations Development Programme (2011) confirms that Africa faces considerable challenges in adapting to the long-term impacts of climate change and extreme climate events. Extreme events have led to severe economic losses over the years and have increase farmers' needs for weather forecast information that will enable predictions beyond the timeline of indigenous knowledge in decision-making processes. Agricultural stakeholders have to contend with the high vulnerability due to low levels of adaptive capacity and poor weather services across many developing economies. Ensuring that decision makers and end users especially smallholders are able to respond to the medium- and-long-term implications of climate change is important in promoting climate-resilient development (Ziervogel and Zermoglio 2009).

Jones and Carabine (2013) observed that the need for medium- and long-term climate information to support climate-resilient development planning in Africa is clear to generate long-term strategies for loss and damage, adaptation finance, climate resilient agricultural practices, and other variants such as climate-compatible development. In the Sub-Sahara Africa, farming systems are highly reliant on prevailing environment, and the existence of traditional knowledge offers elements of resilience (FAO 2011).

In managing the adverse impacts of extreme climate events and climate variability, local farmers have to adjust to harsh weather conditions by relying on weather forecast services which would help moderate potential damage and adverse consequences. The accessibility and usefulness of weather information are therefore critical factors that will affect farmer's ability to adapt (Alston 2013; Serna 2011; Bryan et al. 2009).

In Nigeria, the agricultural production system is predominantly rain-fed, hence highly vulnerable to seasonal variability which affects agricultural livelihood of smallholders and landless laborers who depend on agriculture, as one of the most vulnerable countries among developing African economies (Vermeulen et al. 2012). The impacts of long droughts and deficit rainfall will be mostly felt by the rural population who represent over 80% of the Nigerian population who have less means to adapt and/or diversify their livelihoods (World Bank 2013). Nigeria may lose about 11% of GDP ($460 billion) to drought risk event by 2020 and about 25% by 2050.

Smallholders usually depend on traditional weather predictions in taking production decisions (Mugabe et al. 2010). A variety of traditional indicators have been developed in taking decisions weather-dependent farm management decisions. Over time, some traditional weather and climate indicators consequently are no longer patterned base on existing climate behavior (Sanni et al. 2012). Variability in weather parameters has resulted into significant changes in agricultural production cycle such that farmers can no longer accurately predict start and duration of rainfall seasons by relying on indigenous knowledge (Awolala 2018; Madzwamuse 2010). Weather information has shown potential for improving resilience of agriculture to climate shocks and uncertainty, as a fundamental basis upon which smallholders will adapt to effects of climate change (Food and Agriculture [FAO] 2013, 2015).

Nigeria plans "Climate-Smart Agriculture" under the Nigeria's National Determined Contributions of COP21 Paris Agreement for which weather and climate information services are foremost safety nets. Adapting to impacts through access to timely weather forecasts and other adaptive mechanisms are foremost in Africa's policy dialogues and socio-economic development agendas (Oyekale 2015).

Policy makers' have thus considered several alternatives for enhancing farmers' adaptive capacity in the face of changing climatic parameters. A national strategy towards climate resilience, therefore, is to strengthen local resilience and adaptive capacity of vulnerable smallholders to facilitate their investments in Climate Smart Agriculture National Adaptation "Action" Plan towards a resilient agricultural sector (IFAD 2017; CCAFS 2015). One of the strongly identified pathways to meet this Action Plan is the utilization of weather and climate information to facilitate smallholder decisions for adaptation planning. Despite the inherent uncertainties that are associated with climate science and weather forecasting in making informed decisions on agricultural investments aimed at optimizing use of scarce resources available, there is no enough evidence of understanding gaps in the uptake of weather forecast information for long-term strategies of climate resilience, particularly in the Sub-Saharan Africa including Nigeria (African Climate Policy Centre [ACPC] 2011).

While studies in developed countries reveal that farmers benefit substantially from using seasonal forecasts (Mjelde et al. 2000), seasonal weather information has nevertheless been found to have very minimal impacts as it remains largely unutilized by smallholder farmers in Nigeria despite facing food deficits (Awolala 2018). This study analyzed major drivers of farmers' use of weather information as basis for adaptation decisions in the savannah area of Ondo State, Southwest of Nigeria.

## Materials and Methods

The study site was the Derived Savannah agro-ecology of Akoko North-West Local Government Area of Ondo State, Nigeria. Akoko North-West is located between latitudes 7° 38′ and 7.6° 27.6′ N and longitudes 5° 47′ and 5.8° 38.4′ E of the equator with an elevation of 486 m in Guinea Savannah agro-ecological zone (accessed from Google maps, 2017). Akoko North-West Local Government Area has an estimated population of 213,792 and covers a total area of 512 km$^2$ (National Population Census 2006). Ondo State is located in the Southwest of Nigeria, entirely within the tropics with tropical climate broadly classified into rainy season (April–October) and dry season (November–March). Temperature throughout the year ranges between 21 °C to 30 °C and humidity is relatively high. The annual rainfall varies from 2,000 mm in the south to 1,150 mm in the north. Ondo State enjoys luxuriant vegetation with high forest zone (rainforest) in the south and subsavannah forest in the northern fringe (Fig. 1).

Cross-sectional data used for this study were obtained using a structured questionnaire administered during a farm field survey of arable food crop farmers. Multistage sampling technique was used to select the respondents. In the first stage, Akoko North-West Local Government Area (LGA) was

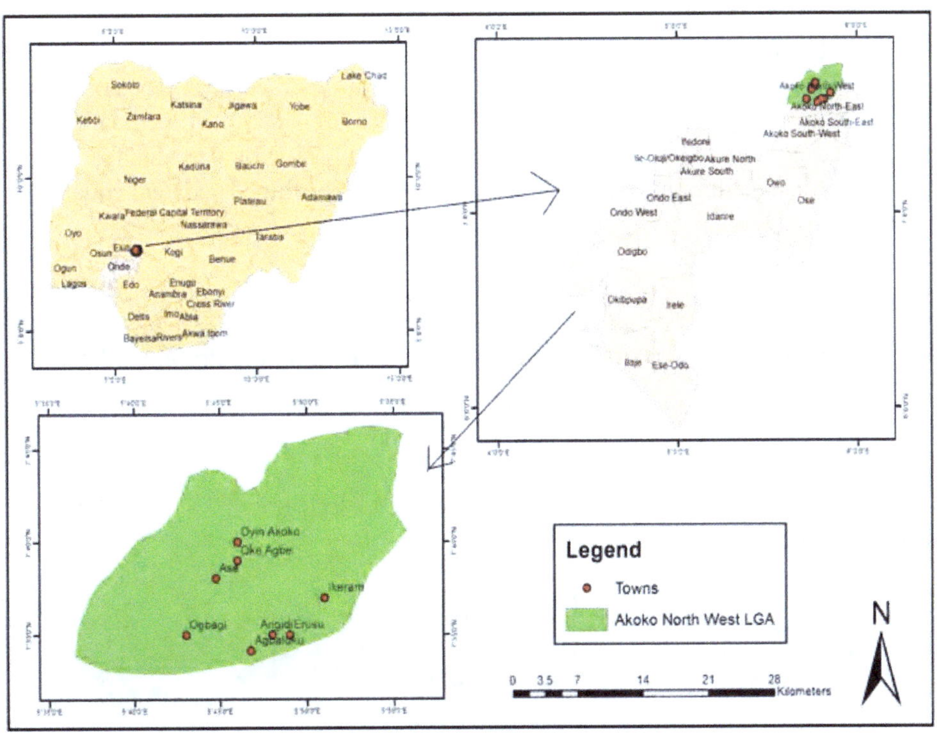


**Fig. 1** Map of Akoko North-West showing selected villages in Ondo State

purposively selected in guinea savannah agro-ecological zone of Ondo State given its dominance of major production of cereals and high dry vegetation. This zone is characterized by low rainfall and long dry period. The savannah area, however, has the potential for producing major food staples in meeting consumption and export requirement towards 2023 National Food Security Agenda in the State. In the second stage, eight farming communities were randomly selected from Akoko North-West LGA using a simple random sampling technique. The final stage involved the random sampling of 20 respondents in each farming community which resulted into a total of a hundred and sixty respondents sampled for the study.

## Econometric Model

### Heckman Two-Step Procedure

Heckman Probit Model was used to Analyse Determinant of Smallholders' Use of Weather Information in decision making processes. When a decision process by farmers to adopt an innovative strategy requires more than one step, models with two-step regressions are usually used to correct for the selection bias generated during the decision-making processes which implies the use of Heckman's Sample Selectivity Probit Model (Heckman 1976; Maddison 2006). For model specification, the first step involved thc analysis of determinants of access to weather forecast information (selection model) and the second step was the smart decision taken, conditional on the first stage of access to weather information (outcome model), adapted from Awolala et al. and Oyekale (2015). The Probit model for sample selection assumes that there exists an underlying relationship. The latent equation is given by:

$$y_j^* = x_j\beta + u_{1j} \tag{1}$$

Such that the binary outcome is only observed given by the Probit model as:

$$y_j^{probit} = \left(y_j^* > 0\right) \tag{2}$$

The dependent variable is observed only if the observation "j" is observed in the selection equation:

$$y_j^{select} = \left(z_j\delta + u_{2j} > 0\right) \tag{3}$$

$$u_1 \approx \mathrm{N}(0, 1)$$
$$u_2 \approx \mathrm{N}(0, 1)$$
$$corr(u, u_2) = \rho$$

where, x is a k- vector of regressors, z is an m-vector of regressors; $u_1$ and $u_2$ are error terms jointly normally distributed, independently of x and z with zero expectations. When $\rho \neq 0$, standard Probit techniques apply to Eq. (3) give biased results. Hence, the Heckman Probit provides consistent, asymptotically efficient estimates for all parameters in such models (StataCorp 2003). Therefore, to analyze the determinant of farmers' utilization of weather forecast information in taking specific adaptations decisions, the marginal effects of a unit change in an independent variable on the probability $P(Z = 1|\ X = x)$ were obtained given that all other variables are held constant. It is mathematically expressed as:

$$\frac{\delta P(Z_i = 1|x_i)}{\delta x_i} = \frac{\delta E(Z_i = 1|x_i)}{\delta x_i} = \varphi(x_i'\beta) \quad (4)$$

Table 1 presents variables that were analyzed in the Heckman two-step econometric model.

**Table 1** Variable list for the econometric analysis

| Variables | Descriptions | Measurements |
|---|---|---|
| *Dependent variables* | | |
| Agricultural smart decisions | Smart agricultural decisions taken by farmers, given weather information received (Yes =1) | Yes = 1; 0 if otherwise |
| Farmer's access to weather information | Farmer has access to weather information | |
| *Independent variables* | | |
| Gender | Gender of farmer | Female = 1; 0 if otherwise |
| Farming experience | Farming experience of household head | Number |
| Farm size | Farm size cultivated | Hectares (ha) |
| Credit | Access to agricultural credit finance | Yes = 1, 0 if otherwise |
| Agricultural extension access | Access to agricultural extension services | Yes = 1, 0 if otherwise |
| Farmers Group membership | Hold membership of Farmers Group | Yes = 1, 0 if otherwise |
| Access to weather forecasts | If farmer receives weather information | Yes = 1, 0 if otherwise |
| *Weather information* | | |
| Rainfall onset date | Rainy season onset date | Yes = 1, 0 if otherwise |
| Cessation date | Rain cessation date | Yes = 1, 0 if otherwise |
| *Agricultural decisions* | | |
| Taking Smart Agricultural decisions | Has taken smart adaptation practices | Yes = 1, 0 if otherwise |

Source: Authors 2017

## Results and Discussions

### Drivers of Farmers' Use of Weather Information for Smart Decisions

The result from the Heckman two-step regression estimates result in Table 2 reveals that access to credit is a major driver increasing farmers' propensity to adapt with climate-smart practices at 5% level of significance while increasing access to information onset date, farm size, access to extension reduce farmers' their ability to take climate-smart actions in the derived savannah area. However, gender and access to agricultural extension service were major drivers positively facilitating access to weather information while farmers' experience negatively influence farmers' access to weather information in the derived savannah area of Ondo State, Nigeria.

Access to credit has a positive coefficient with the adoption of climate-smart practices ($p<0.05$). An increase in access to credit will increase the likelihood that farmers would adopt climate-smart adaptation practices conditional on access to weather forecast information. It is viewed that with credit facilities, interested farmers will be able to bear the cost associated with climate smart practices. For instance, when there is a forecast that rainfall is likely to be unusually low, farmers can grow crops that are drought resistant and sensitive to water availability

**Table 2** Heckman estimates of major drivers of farmers' use of weather information

| Variables | Coefficient | Standard error | Z-value | \|P>z\| |
|---|---|---|---|---|
| Taken smart agricultural decisions | | | | |
| Access to rainfall onset date | −0.2677499 | 0.1405112 | −1.91 | 0.057[a] |
| Access to rainfall cessation date | 0.1754594 | 0.1307748 | 1.34 | 0.180 |
| Farm size in hectares | −0.0765449 | 0.0403879 | −1.90 | 0.058[a] |
| Access to agricultural extension service | −0.0335072 | 0.3270912 | −0.10 | 0.918 |
| Access to credit | 0.2943015 | 0.1500301 | 1.96 | 0.050[b] |
| Constant | 0.6239116 | 0.5747793 | 1.09 | 0.0278[b] |
| Access to weather information | | | | |
| Gender | 0.9495208 | 0.2676775 | 3.55 | 0.000[b] |
| Farming experience | −0.0417005 | 0.0096622 | −4.34 | 0.000[b] |
| Access to agricultural extension service | 0.736214 | 0.3857535 | 1.91 | 0.056[a] |
| Constant | 0.3007832 | 0.4244483 | 0.71 | 0.479 |
| Total observations | 160 | | | |
| Censored obs | 99 | Uncensored obs | 61 | |
| Prob > $chi^2$ | 0.0621 | | | |

Source: Field Survey 2017
[a]Statistically significant at 0.05
[b]Statistically significant at 0.1

to improve profit unlike farmers who also received forecast information but lack access to credit. Therefore, credit facility should be strengthened for smallholders' adoption of climate-resilient technology adaptation actions in the derived area.

Seasonal rainfall onset date and farm size both have negative coefficients with the adoption of climate-smart adaptation practices ($p<0.05$). An increase in access to seasonal rainfall onset date, likewise increase in farm size, will reduce the likelihood that farmers would adopt climate-smart adaptation practices conditional on access to weather forecast information. These are important policy drivers that should be addresses because they signal danger to farmers' adaptation with climate-smart practices conditional on access to weather forecast information. There are certain socioeconomic and cultural factors which might be responsible for such findings. It is also believed that medium and large-scale farmers might be reluctant in taking climate-smart decisions due to the huge costs involved.

The positive coefficient of gender and access to extension service are both significant at 5% level ($p<0.05$). The result implies that male farmers are more likely to have access to scientific weather forecast information unlike their female counterparts. It thus becomes obvious that this male farming population should be given attention as major pathway to ensure farmers' resilience with regards to utilization of scientific weather forecast information.

Likewise, increase in access to agricultural extension service, it will be more likely that farmers would have access to weather forecast information in the study area. This is because the role of agricultural extension is critical to move weather forecast information from weather service providers especially the Nigerian Meteorological Agency (NiMet) and other weather forecast agencies in other to ensure a return on investment by translating new knowledge into innovative practices which will enhance farmers' resilience.

Further results explained the negative coefficient of farmers' experience with access to weather forecast information ($p<0.05$). As farmers' experiences increase, there is a declining likelihood that they will have access to scientific weather forecast information. This might be attributed to their over reliance on traditional experience and knowledge of monitoring weather with local indicators in the study area. This is another important policy variable indicating that there should be improvements in farmers literacy and awareness campaign to encourage the use of scientific-based weather forecasts information to help facilitate climate-smart adaptation actions other than traditional forecasting techniques in the dryland. Table 2 presents the econometric results of factors influencing the use of weather information for smart decisions in the study area.

## Conclusions and Policy Recommendations

Access to rainfall cessation date has a positive significant effect on farmers' use of improved weather information. Increasing access to seasonal rain probabilities forecast through information sharing increases the likelihood that farmers will use improved weather information for smart production decisions. Weather services such as rainfall forecasts and advisory planting calendars based on length of growing season should be location specific and timely communicated by the Nigerian Meteorological Agency (NiMet) in a mode that is understood at farmers' level. Such information on rainfall cessation date is considered significantly helpful for farmers adaptation planning and resilience.

Credit also significantly facilitates farmers' use of weather information in taking climate-smart decisions. It is viewed that with credit facilities, interested farmers will strengthen their adaptive capacity to afford whatever transaction cost that might be associated with climate smart practices if they have adequate financial assistance. Better incentive-based credit facility should be developed as a farmer-oriented financial product to strengthen smallholders' decisions for climate-resilient technology actions in the dry land area. Similarly, Policy makers should ensure functioning financial institutional services especially rural access to bank credits and informal credits are available to help farmer's withstand unforeseen circumstances that may arise from investing in innovative climate smart practices.

In terms of access, male farmers were found to have more access to weather information. Agricultural decisions are traditionally considered as responsibility of the household head, and most of the times, it is the male gender. Therefore, weather information utilization could be more guaranteed when male farmers are used as an entry point to allow for quick and wider coverage among farmers given their strong influence as decision makers. Policy makers and agricultural stakeholders at the federal, state, and local government levels must ensure that rural education is provided for male household heads to fast-track local uptake of improved weather information as weather risk adaptation strategy.

Better access to agricultural extension services also increase farmers' likelihood to have access to weather forecasts. Agricultural Extension Department at the Ministry of Agriculture and Rural Development must ensure that periodic extension education and training services serve as opportunities to share updated information on weather with farmers and provide advisory services on smart agro-weather management technologies. A functional and effective extension service system has great potential in facilitating early action on weather risk management through climate smart practices by the increasing farming population in the derived savannah area of Ondo State, Nigeria.

These drivers should be considered for a review as strategies for implementing the National Climate Smart Agriculture Adaptation Plan (NCSAAP) in Ondo State, Nigeria, as stated in the revised National Agricultural Policy which targets agricultural livelihood resilience towards meeting the 2030 National Food Security Agenda.

## References

African Climate Policy Center [ACPC] (2011) Assessment of Africa's Climatic records and recording networks including strategic for rescuing of climatic data. Working Paper 3 of the African Climate Policy Center (ACPC) Of United Nations Economic Commission for Africa (UNECA) under the Climate for development in Africa (Climdev Africa) programme

Alston M (2013) Gender mainstreaming and climate change. Women's Stud Int Forum 47:1–8

Awolala DO (2018) Climate smart adaptations in the African tropics: scaling weather information for decision support outcomes in Nigeria savannahs. In: Leal Filho W (ed) Handbook of climate change resilience. Springer, Cham, pp 1–33. https://doi.org/10.1007/978-3-319-71025-9_169-1. Online ISBN 978-3-319-71025-9

Bryan E, Deressa T, Gbetibouo GA, Ringler C (2009) Adaptation to climate change in Ethiopia and South Africa: options and constraints. Environ Sci Pol 12:413–426

Climate Change, Agriculture and Food Security [CCAFS] (2015) Implementing climate-smart agriculture for enhanced food security and resilience. Progress report led by the International Center for Tropical Agriculture (CIAT) and other Research Centres

FAO (2011) Climate change and food systems resilience in sub-saharan Africa (eds: Ching LL, Edwards S, Scialabba NE-H). Office of Knowledge Exchange, Research and Extension, FAO, Rome

FAO (2013) Climate smart agriculture source book. Food and Agriculture Organization of the United Nations, Rome

FAO (2015) Barriers, incentives and benefits in the adoption of climate smart agriculture: enhancing agricultural mitigation within East Africa Dairy Development project (EADD) in Kenya. FAO, Rome

Heckman JJ (1976) The common structure of statistical models of truncation, sample selection and limited dependent variables and a simple estimator for such models. Ann Econ Soc Meas 5:475–492

IFAD (2017) IFAD Strategy and Action Plan on Environment and Climate Change: 2019–2025. Document: EB 2018/125/R.12.

Jones L, Carabine E (2013) Exploring political and socio-economic drivers of transformational climate policy: early insights from the design of Ethiopia's Climate Resilient Green Economy strategy. Research reports and studies. Overseas Development Institute, London

Lal R, Kraybill D, Hansen DO, Singh BR, Eik LO (2016) Research and Development Priorities. In: Lal R, Kraybill D, Hansen D, Singh B, Mosogoya T, Eik L (eds) Climate Change and Multi-Dimensional Sustainability in African Agriculture. Springer, Cham. https://doi.org/10.1007/978-3-319-41238-2

Maddison D (2006) The perception of and adaptation to climate change in Africa. Discussion Paper No. 10. Centre for Environmental Economics and Policy in Africa (CEEPA). University of Pretoria, Pretoria

Madzwamuse M (2010) Climate governance in Africa: adaptation strategies and institutions. Heinrich Böll Stiftung (HBS), Nairobi

Mjelde et al (2000) Climate change and multi-dimensional sustainability in African agriculture

Mugabe FT, Mubaya CP, Nanja DH, Munodawafa A, Gondwe P, Mutswangwa EA, Changonda C, Masere P, Makuvaro V, Dimes J, Murewi C (2010) Using indigenous knowledge for climate adaptation in Southern Zambia and South-Western Zimbabwe. Zimb J Sustain Dev Afr 11(2):237–256

NPC (2006) National Population Census Report, Nigeria

Oyekale AS (2015) Factors explaining farm households' access to and utilization of extreme climate forecasts in Sub-Saharan Africa (SSA). Environ Econ 6(1): 91–103

Sanni et al (2012) Traditional capacity for weather prediction, variability and coping strategies in the front line states of Nigeria. Agric Sci 3:625

Serna JM (2011) Drought Assessment Northern Eastern Kenya (Wajir East, South and Mandera). Northern Eastern Kenya Drought Assessment Report

StataCorp (2003) Stata base reference manual. Volume 4, G – M, release 8. Stata Corporation, College Station

UNDP (2011) Human Development Report 2011. Sustainability and equity: a better future for all. Palgrave Macmillan, New York

UNEP (2018) United Nations Environment Annual Report

Vermeulen S, Zougmore R, Wollenberg E, Thornton P, Nelson G, Kristjanson P, Kinyangi J, Jarvis A, Hansen J, Challinor A, Campbell B, Aggarwal P (2012) Climate change, agriculture and food security: a global partnership to link research and action for low-income agricultural producers and consumers. Curr Opin Environ Sustain 4:128–133

World Bank (2013) Weather and climate resilience: effective preparedness through National Meteorological and Hydrological Services (NMHSs) in the World Bank Development Report of Rogers DP, Tsirkunov VV (2013) Directions in development – environment and sustainable development. World Bank, Washington, DC

Ziervogel G, Zermoglio F (2009) Climate change scenarios and the development of adaptation strategies in Africa: challenges and opportunities. Clim Res 40(2–3):133–146. https://doi.org/10.3354/cr00804

7

# Panel Analysis of the Relationship Between Weather Variability and Sectoral Output in Kenya

Olga Nekesa Mulama and Caroline Wanjiru Kariuki

## Contents

Introduction .
Theoretical and Empirical Context .
  Theoretical Context
  Empirical Context
  Interactions Between the Variables Used in This Research ...
Approach to Identify the Relationship Between Weather Variability and Sectoral Output .
  Data Collected
  Model on the Relationship Between Weather Variability and Sectoral Output
  Procedure ...
Establishing the Relationship Between Weather Variability and Sectoral Output
  Unit Root Test
  Long-Run Relationship Between Temperature and Sectoral Output
  Effect of Weather Variability on Sectoral Output
Conclusion
References

**Abstract**

Climate change and economic growth are closely connected. Climate change has the potential to reduce economic growth in developing countries due to their limited ability to respond to the negative impacts of a changing climate. A better understanding of weather variability can enhance climate change policies, which

O. N. Mulama · C. W. Kariuki (✉)
Strathmore Institute of Mathematical Sciences, Nairobi, Kenya
e-mail: olga.mulama@strathmore.edu; cwkariuki@strathmore.edu

would help to support economic growth in these countries. As such, this research sought to examine if there is a long-run relationship between sectoral output and weather variables (temperature and rainfall) and to analyze the effect of weather variability on sectoral output using a panel of 13 sectors in Kenya.

A Pedroni cointegration test was carried out to find out if there exists a long-run relationship among the variables and thereafter, a fully modified ordinary least squares regression was conducted to establish the effect of weather variability on sectoral output. The results indicate that there is a long-run relationship between temperature and sectoral output. Moreover, temperature has a larger effect on sectoral output compared to rainfall. With the evidence gathered from this research, it can be concluded that weather variability has an economic effect on sectoral output in Kenya. Given this, the Kenyan government needs to take a keen interest in understanding the effect of weather variability on the economy and in the broader picture, take steps to mitigate climate change.

**Keywords**

Weather variability · Climate change · Temperature · Rainfall · Sectoral output · Economic growth · Kenya · Panel data · Cointegration · Fully modified OLS

## Introduction

Weather refers to the temporary conditions of the atmosphere, which is the layer of air that surrounds the earth. There are various factors that can change the atmosphere such as temperature, air pressure, humidity, wind speed, wind direction, altitude, and topography, among others. Together, these factors determine the weather in a certain area (National Centers for Environmental Information 2018). On the other hand, climate refers to the state of the weather over a long period of time in a specific area. Predictable weather is important for social and economic planning as it results in sustainable development. However, climate change could affect health, infrastructure, transportation systems, as well as energy and food supplies. There is also a global challenge of acute water scarcity, which can be attributed to the extreme temperatures. On a global scale, there is a cry for environmental conservation since extreme weather occurrences have been found to be directly linked to the environment's state. Through their research, Mortimore et al. (2009) provide insight on the pattern that developing countries have recorded in terms of extreme weather occurrences. The authors find that there has not been much investment and development to aid in preventing the damages that come with extreme weather occurrences in these countries. Developing countries are currently faced with considerable development constraints and to top it off, their poor coping and adaptation strategies to a changing environment push them deeper into development issues. The agricultural, livestock, fisheries, tourism, energy, forestry, health, and water sectors usually take the biggest blow when extreme weather events take place due to their relation and dependent nature on weather. The labor force and industries are also affected due to extreme weather events and this eventually ends up having an adverse effect on the peace and stability of a country.

In Kenya, the country's wildlife and other tourist attractions are vulnerable to the impact of climate change. Ferrara's (2015) research revealed that tourism in Kenya would be significantly affected by a change in wildlife migration patterns (e.g., wildebeest migration) caused by changing weather. Consequently, this would have a negative effect on the country's economy given that tourism is one of the major contributors to Kenya's Gross Domestic Product (GDP). Mount Kenya, which is another major tourist attraction, has snowcaps that are melting due to increased global temperatures. Additionally, the Kenyan economy is highly dependent on agriculture which contributes approximately 33% to the nation's GDP. The agricultural sector in Kenya employs about 40% of the total population with more than 70% coming from rural areas (Food and Agriculture Organization 2019). The Food and Agriculture Organization ( 2019) also reports that farmers who are used to rain-fed farming systems are being pushed to areas that are vulnerable to drought due to increased human population. Furthermore, the unpredictability of weather patterns resulting from climate change does not favor their farming practices. In many developing countries, households face difficulty in adapting to climate change and very few households make a step and change their farming practices in response to climate change. Okoba et al. (2013) argue that despite the possibility to change this, many households are unable to make the costly investment and join the bandwagon in the agricultural shift. Despite developing countries having information on extreme weather occurrences, the impact of the occurrence of an extreme weather event is still very high. News of extreme weather patterns across Africa is increasingly becoming frequent and policymakers can no longer overlook this effect, which is a result of climate change. As temperatures continue to rise as a result of global warming, there are bound to be water shortages in Africa causing droughts and in the short term, occurrences of flash floods as well. The cycle effect is that agricultural output will be affected thus causing a drop in rural incomes for farmers who depend on agriculture and resulting in a decline in consumption and investment.

Evidently, climate and agriculture are highly correlated as discussed by Linke et al. (2015) who carried out a research on the relationship between climate and social instability in Kenya. Agriculture is highly dependent on favorable weather conditions. Low output from farming during unfavorable weather conditions would result in low returns on the farmers' investment and in the next agricultural season, farmers may not have enough capital to invest for their next harvest. This is a vicious cycle whereby output that is affected by the weather would affect levels of income and this in turn affects the input necessary to produce more and the cycle continues. Dell et al. (2012) seek to bring out the evidence that in fact weather and economic variables do have a direct relationship. Their research examined the relationship between temperature shocks and economic growth in both rich and poor countries. Their findings show that higher temperatures reduce agricultural output, industrial output, and political stability. Not only do changes in the weather affect the agricultural and industrial sectors of a country, but the health of the population as well. The general health of a population is usually at risk in cases of extreme weather occurrences like drought and floods. In Kenya, infectious disease outbreaks such as waterborne, rodent-borne, and vector-borne diseases have been associated with flooding especially in areas with poor drainage systems (Okaka and Odhiambo 2018). Moreover, being on the equator, the temperate climate provides a perfect breeding ground for mosquitoes that spread malaria. Yanda and Mubaya (2011) report a rise in mosquito population in the occurrence of erratic weather where the spread of malaria and other waterborne diseases causes Kenya's income to shrink in affected areas. This is because the current labor force must attend to their sick family members and current and/or future labor force could also decrease should they succumb to diseases. Furthermore, the World Health Organization (2018) reports that diseases transmitted by insects are emerging and reemerging and epidemics are reported more frequently than before in Africa. Evidence points to climate change being a major contributor to this public health crisis. With the weak adaptive mechanisms in developing nations, adverse weather patterns are bound to have a huge impact on the entire economy affecting the majority of the population. Thus, there is a need for policymakers to plan accordingly and most especially conserve the environment and thus the economy for future generations. The erratic nature of weather patterns without adequate adaptation mechanisms has cost Kenya much more than it would have cost the country to put in place preventative measures that would reduce the scale of damage. For example, During the El Niño that occurred in 1997, Kenya made losses of up to USD one billion (Ngecŭ and Mathu 1999). The lack of preparation in the country adversely affected physical infrastructure due to the heavy downpour and further affected road, air, and rail transport networks.

Given this, Nyangena (2016) studied the impact of weather variability (rainfall and temperature) on economic growth in Kenya. Nyangena (2016) cited the need for further empirical explorations and as such, this research analyzed the effect of weather (temperature and rainfall) variability on sectoral output in Kenya using panel data. Specifically, this research sought to establish if there is a long-run relationship between temperature, rainfall, and sectoral output in Kenya and to determine the effect of temperature and rainfall on sectoral output in Kenya. The sectors captured in this research were agriculture and forestry; mining and quarrying; manufacturing; electricity and water; construction; wholesale and retail trade; hotels and restaurants; transport and communication; financial intermediation; real estate, renting, and business services; public administration; education; and other services.

It is important to remember the past so as to model better decisions for the future. Consequently, this chapter will make a contribution to literature by providing a deeper understanding of the effect of weather variability on economic output in Kenya using historical data. This will ultimately aid policymakers in planning and making decisions that will have a positive effect on the welfare of future generations. More specifically, the findings from this research will provide further insights on the effect of weather (temperature and rainfall) variability on sectors that contribute to Kenya's economic growth. Further, the results will assist environmentalists in campaigning for environmental conservation and promote the use of environmentally friendly products, methods, and clean energy. Institutions such as The National Environment Management Authority of Kenya and the Green Belt Movement together with the Ministry of Environment and Natural Resources and are not only involved in environmental campaigns but also advocate for policies and strategies that will help conserve the environment and mitigate the effects of climate change. The Ministry of Agriculture also stands to make more informed decisions that will benefit farmers and offer insight on policies regarding food security in Kenya related to weather variability.

## Theoretical and Empirical Context

### Theoretical Context

#### Chaos Theory

Making future weather forecasts is hard. It is only predictable to forecast the weather behavior in the future when a trend is observed from some initial weather conditions. Stephens et al. (2012) state that it is easier to predict climate than weather because weather is governed by an imbalance of the world's energy forces. A chaotic system is purely deterministic assuming absolute dynamism to the extent of exhibiting unpredictability in standard statistical tests, rendering it chaotic. Chaos theory, as described by Lorenz (1963) is the confusion, disorder, unpredictability, turbulence, and uncertainty in a spectrum of things. A chaotic system is one that is affected by initial occurrences and dynamic nonlinear patterns that are not fundamentally random. Understanding the initial points is fundamental for predictability to hold. Weather patterns are nearly impossible to predict due to their chaotic nature. However, with the advent of computers and their ability to handle massive amounts of data, there is a better possibility of predicting the weather. An essential observation would be to see how many times the weather predictions have merged with the actual weather occurrences. As observed from past weather occurrences, there is generally a disparity most times and this is attributed to the possibility of climate change taking place. A general standard stochastic test can be useful in distinguishing a set of randomness and chaos.

#### Climate Change Theories

Hulme (2009) describes climate science as a function of the limits of science itself. Mathematicians working on climate systems find that uncertainties in climate prediction arise not only from initial conditions and forcing scenarios but also from model formulation. Further, the anthropogenic global warming theory admits that the climate system is naturally chaotic. Most people are familiar with anthropogenic global warming, which is manmade global warming. Under the theory of anthropogenic global warming, it holds that manmade greenhouse gases are the main cause of global warming that has occurred during the past 50 years. As the years have progressed, loopholes have been found in this theory and new evidence shows that other natural factors contribute to modern global warming. The theory of anthropogenic global warming is contested by other theories, some of which include the theory of the Bio-thermostat, which claims negative feedback from biological and chemical processes offset whatever positive feedback might be caused by the rising carbon dioxide levels (Idso and Singer 2009). Sud et al. (1999) observed cloud coverage and formation. They found that the cloud coverage acted as a thermostat to regulate sea surface temperatures to a temperature of 28 °C and 30 °C. In a separate theory on human forcing, Matsui and Pielke (2006) explain that human activities such as deforestation, aerosols and ozone, coastal development, and jet contrails (the trail left behind by jets) transform the earth's surface. Additionally, carbon dioxide emissions as a result of human activities play a dominant role on long-term climate change. The US Global Change Research Program (2014) reports changes in the earth's climate due to the earth warming, with a key indicator of climate change being extreme weather and climate events.

### Economic Growth

The neoclassical theory on economic growth states that the stock of capital, supply of labor, and technological developments are factors that affect the growth of an economy. The theory further explains the law of diminishing returns where increasing capital and labor have a limited impact on increasing economic growth. The new economic growth theories criticized the neoclassical theories stating that labor productivity does not have diminishing returns but rather has increasing returns and emphasis is put on the type of capital investment. Smith (1776) emphasized the dependency of economic growth on labor efficiency through division of labor. Additionally, Montesquieu and De Secondat (1748) argued that excess heat made men slothful and dispirited. As such, adverse climatic conditions affect laborers' well-being and this in turn affects their output ultimately affecting economic growth. Moreover, the Malthus theory predicts that economic growth may have limitations caused by changes in the climate, overpopulation, and scarcity of resources.

In his earlier studies, Barro (1999) explains that the growth rate of real per capita GDP was associated with higher levels of schooling. This is because an additional year in school enhances workforce capability to work and increases productivity. Real GDP per capita as used in this context is the measure of the economy's total output in a country divided by the population and further adjusted for inflation. Workers level of education is a measure of human capital and is directly related to productivity. Under the Harrod–Domar theory, capital formation is given more weight on its effect on economic growth. The Harrod–Domar theory emphasizes the importance of national savings and productivity of capital investment as central to economic growth. The growth of an economy is dependent on demand created from newly generated income such that output produced by new investments can be fully absorbed. Ultimately, this should increase the production capacity of the economy. This theory further assumes that labor and capital are complementary to each other.

## Empirical Context

A report by the Intergovernmental Panel on Climate Change (2014) warned that global temperatures are expected to increase by an estimate of 4 °C by 2100 (World Bank

2012). Considering these projections, economists and policymakers are faced with the need to quantify the impact of weather patterns, specifically rising temperatures on economic activity. Nordhaus (2010) cites the significant effects of extreme weather events on the level of the US GDP. According to Dell et al. (2012), in their analysis of the relationship between temperatures and aggregate economic activity, hot countries tend to be poor, with national income falling 8.5% per degree Celsius in the world cross-section. With the recent interest in climate change, it is a matter of concern to understand the impact of the weather on all aspects of life. Colacito et al. (2019) estimated the effect of temperature on the growth rate of US GDP using a panel of the US states. From their estimates, they concluded that if the current trend in rising temperatures does not change, then a drop of economic growth by up to one third could occur. Further, in their estimation, they found that summer and fall exhibited opposite signs. Summer and fall had opposite effects on the US Gross State Product (GSP). This meant that rising temperatures during summer decreased economic growth as estimated by GSP and a significant rise during fall as the temperatures decreased. There has been a difficulty in concluding the general effect of weather patterns using annual data as each year presents different seasons. This heterogeneous effect can be dealt with by using more frequent data to represent the weather patterns.

In their inquiry, Linke et al. (2015) speculated that the future will get warmer, wetter, and wilder climates will be experienced. Ultimately, there is and still will be a surge of migrants and new wars. Intercommunity battles for resources increase conflicts and violence driven by calculation and political gain. According to the Food and Agriculture Organization of the United Nations (2019), increased competition for natural resources in Kenya has sparked escalated conflicts in some of the areas. Abidoye and Odusola (2015) examined the link between economic growth and climate change and found that the vulnerability of most African economies to climate change is of concern. A fair majority of the working population largely depend on informal employment in economic sectors that are sensitive to climate change (such as agriculture, forestry, energy, tourism, coastal, and water resources). Temperature was seen to have a negative impact on GDP growth where a 1 °C change in temperature would result in an inverse change in GDP growth by 0.67%. Given this, there is a need to manage the impact of climate change on Africa's economy to ensure developmental growth. In another article, Deryugina and Hsiang (2014) explored within-country variation in the United States and found that daily temperature above certain thresholds (15 °C and 30 °C) reduces the productivity level. In a related paper, Zivin and Neidell (2014) find that warmer temperatures reduce labor supply in the United States, thus providing an economic rationale for why climate change might affect economic activity in a developed economy. There has been a correlation between poverty and heat as suggested by Montesquieu and De Secondat (1748), who argue that excess heat made men slothful and dispirited.

Several other studies have focused on agriculture specifically and there is an evident inverse relationship between agricultural productivity and extreme weather occurrences although this is dependent on the crop type. Beyond agricultural constraints as a result of climate change, Awuor et al. (2008) found that floods lead to serious loss of life and property, which coastal towns such as Mombasa face as a result of climate change. Mombasa is a major contributor to Kenya's GDP through tourism, industries, imports, and exports and serves both Kenya and landlocked countries around it. Floods lead to loss of life, displacement of people, and destruction of infrastructure. Floods can occur at any time despite weather or climate patterns influencing when and where these floods occur (Smith 2002). Although drought affects more people than floods do, disease outbreaks such as malaria, typhoid, bilharzia, and dysentery have been seen to cripple labor productivity. Flood disasters are associated with the presence of flood hazards due to the nature of the terrain or flood vulnerability, which is attributed to poor regulation, poor land use, degradation of water catchment areas, poverty, population pressures, and settlement patterns (Opere 2013).

In a more recent research done in Kenya, Nyangena (2016) explored the relationship between GDP per capita, annual temperature changes, and annual rainfall and found that a unit change in temperature resulted in approximately 142 times change in GDP per capita arguing that temperature has a greater effect on GDP than rainfall. The empirical literature in Kenya revealed some inconsistencies with other bodies of literature that have sought out to explore the role of weather and climate changes in economic growth (Nyangena 2016). The findings of earlier studies revealed that temperature variability had an inverse relationship with economic growth (e.g., Colacito et al. 2019; Dell et al. 2012) whereas Nyangena (2016) found a positive relationship between temperature and economic growth. Nyangena (2016) determined the long-run relationship between economic growth and weather and used a vector error correction model to analyze the impact of weather variability on economic growth. On the other hand, this research aims to analyze the effect of weather variability on 13 different sectors in the Kenyan economy as opposed to the economy at an aggregate level. The sectors included in this research are: agriculture and forestry; mining and quarrying; manufacturing; electricity and water; construction; wholesale and retail trade; hotels and restaurants; transport and communication; financial intermediation; real estate, renting, and business services; public administration; education; and other services. Additionally, this chapter will also provide insight as to whether there is a long-run relationship between weather variability and output from the 13 sectors that contribute to Kenya's overall economic growth.

## Interactions Between the Variables Used in This Research

The dependent variable in this research was sectoral output (measured using GDP per sector) focusing on the following sectors: agriculture and forestry; mining and quarrying; manufacturing; electricity and water; construction; wholesale and retail trade; hotels and restaurants; transport and communication; financial intermediation; real estate, renting and business services; public administration; education; and other services. The sectors were chosen based on availability of historical information. GDP per sector is the contribution that the sector has on overall GDP. The independent variables were temperature, rainfall, gross fixed capital formation, labor, and the business regulatory environment captured using the country policy and institutional assessment (CPIA) (Fig. 1).

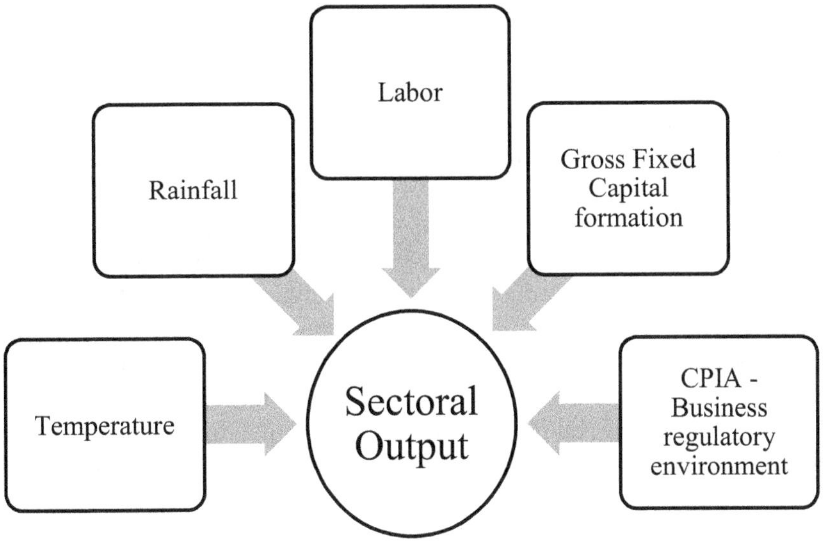


**Fig. 1** Interactions between the variables used in this research

The Cobb–Douglas function drives the concept behind the variables chosen. Labor and capital should have a positive relationship with output, and in this case, the sectoral output is expected to have a positive relationship with gross capital formation because increased investments are expected to increase production and as such there would be more output, which is plowed back to investments and the cycle continues. The CPIA rating is expected to have a positive relationship with sectoral output where policy improvements would positively impact output growth. Labor is also expected to have a positive relationship with sectoral output because the labor force is key to increasing productivity up to a certain level. An increase in productivity due to labor just like capital formation should result in an increase in output, which would be eventually plowed back as investments, which could take the form of increased labor opportunities. The weather variables (rainfall and temperature) on the other hand are expected to have an inverse relationship with sectoral output should there be a significant increase or decrease. For example, in the case of agriculture, a significant increase or decrease in temperature and/or rainfall will affect agricultural yield, reducing production and in the long-run causing a decline in output, which is GDP per sector in this case.

## Approach to Identify the Relationship Between Weather Variability and Sectoral Output

### Data Collected

This research sought to determine the effect of temperature and rainfall on sectoral output in Kenya and further establish if there is a long-run relationship between temperature, rainfall, and sectoral output in Kenya. This research studied the sample overtime using past information, which made it possible to establish the direction and magnitude of the effect of the independent variables on the dependent variable.

According to the data availability on Kenya National Bureau of Statistics, the Kenyan economy has 13 economic sectors. The major industries include agriculture, forestry, fishing, mining, manufacturing, energy, tourism, and financial services. For this chapter, data on all 13 economic sectors over a period of 10 years, from 2008 to 2017 was used to establish the effect of weather variability on the economy. To capture the effects of weather variability of economic sectors in Kenya, the start year of the research was 2008 but due to the unavailability of the most recent data on weather in Kenya, this research was limited to the years 2008–2017. Weather is generally described using the following atmospheric conditions: Temperature, wind, precipitation, and sunlight. Being in the tropics (at the equator) precipitation is generally experienced as rainfall and as such, due to limited data, the variables used to measure weather variability in this chapter are rainfall and temperature over a period of 10 years from 2008 to 2017.

The data used was collected from secondary sources and comprised observations of variables over time with a cross-sectional dimension based on economic sectors. As such, panel data was used for this research. Variables used were temperature in degrees Celsius, rainfall in millimeters, GDP per sector in local currency, gross fixed capital formation in local currency, labor, and CPIA. The data for temperature in degrees Celsius, rainfall in millimeters were obtained from the World Bank Climate Change portal. Data for GDP per sector in local currency were obtained from the Kenya National Bureau of Statistics. Gross fixed capital formation data in local currency, which was measured as the annual net increase in physical assets was obtained from the World Bank Development Indicators. The data for labor, which measures the number of people employed was also obtained from the World Bank Development Indicators, while the CPIA data, which measures the implication of the legal, regulatory, and policy environments that help or hinder private businesses in investing, creating jobs, and becoming more productive was obtained from the World Bank Country Policy and Institutional Assessment database. All variables used for this research had an annual frequency.

## Model on the Relationship Between Weather Variability and Sectoral Output

A panel data model was estimated and used to analyze the data. This model was used to provide information on the effect of temperature and rainfall on sectoral output in Kenya and was estimated using the fully modified ordinary least squares (FMOLS) method. All the variables were transformed into natural logarithms thus allowing the coefficient estimates to be interpreted as elasticities. The dependent variable of the model was GDP per sector (sectoral output) defined as the share that each sector contributes toward GDP. The independent variables were temperature measured in degrees Celsius and rainfall in millimeters. Additionally, following the Cobb-Douglas function of measuring productivity, the gross fixed capital formation was included to account for capital and labor force was used to account for human capital. The business regulatory environment was also captured using the country policy and institutional assessment (CPIA). Below is the econometric model:

$$\begin{aligned} lnGDP\ per\ sector_{it} = \beta_0 + \beta_1 lnCapital\ formation_{it} + \beta_2 lnLabor_{it} + \beta_3 lnTemp_{it} \\ + \beta_4 lnRain_{it} + \beta_5 lnCPIA_{it} + \omega_{it} \end{aligned}$$

Where: $\beta_0$ is the intercept, which is the same for all the sectors and over time. $\beta_1$ to $\beta_5$ are slope parameters. The subscript $i$ refers to an individual sector and $t$ refers to the years from 2008 to 2017. $\omega_{it} = \varepsilon_i + u_{it}$. $\varepsilon_i$ measures the random deviation of each sector's intercept from the common intercept term $\beta_0$ and $u_{it}$ is the residual term for sector $i$ at time $t$. *Capital formation* is the net increase in physical assets, *Labor* captures the number of people employed, *Temp* captures temperature in degrees Celsius, *Rain* captures rainfall in millimeters, and *CPIA* captures the business regulatory environment as a rating (1 = low to 6 = high).

### Procedure

Before investigating the existence of a long-run relationship between weather variability and sectoral output, a unit root test was carried out. Unit root tests are used to find out if a series of variables is stationary or nonstationary. When a series is stationary, the variables are time-invariant and are better suited for making forecasts. This research employed the Im, Peseran, and Shin (IPS) unit root test. Next, a test for the existence of a long-run relationship between weather variability and sectoral output using a cointegration test was carried out. A panel cointegration approach was employed and the principal idea was to investigate the existence of a long-run relationship among the nonstationary variables. Finally, a fully modified ordinary least squares (FMOLS) procedure was adopted to estimate the effect of weather variability on sectoral output. Pedroni (1996) brings forth the FMOLS as a solution to tackling non-exogeneity and serial correlation problems. Additionally, FMOLS addresses the problem of nonstationary variables as well as simultaneity biases.

## Establishing the Relationship Between Weather Variability and Sectoral Output

### Unit Root Test

In order to establish whether the distribution of the variables used in this research does not change with time (stationarity), unit root tests were carried out. The results in Table 1 are a result of conducting an Im, Pesaran, and Shin (IPS) panel unit root test at level, first difference, and second difference. When conducting a unit root test, the null hypothesis is the presence of a unit root and rejecting the null means that the variable is stationary. The null hypothesis is rejected when the p-value is smaller than

**Table 1** Unit root test results

| Variable | Level | First-order difference | Second-order difference |
|---|---|---|---|
| GDP per sector | −0.4362 (1.000) | −2.8004[a] (0.0001) | |
| Temperature | 0.2327 (1.000) | −2.4535 (0.2990) | −2.3519[a] (0.0022) |
| Rainfall | −3.8568[a] (0.0000) | | |
| Capital formation | −1.5808 (0.1887) | −3.2164[a] (0.0000) | |
| Labor | −0.8516 (0.9791) | −0.5463 (0.9995) | −3.0059[a] (0.0000) |
| CPIA | −2.3886[a] (0.0010) | | |

P-values in parenthesis
[a]Statistically significant at the 5% level of significance

**Table 2** Panel cointegration results

| Test | Level |
|---|---|
| Modified Phillips-Perron test | 3.0714[a] (0.0011) |
| Phillips-Perron | 1.9829[a] (0.0237) |

P-values in parenthesis
[a]Statistically significant at the 5% level of significance

0.05 ($p < 0.05$). At level, the null hypothesis is rejected for rainfall and CPIA indicating stationarity. At first difference, the null hypothesis is rejected for GDP per sector and capital formation and finally, at second difference, the null hypothesis is rejected for temperature and labor.

## Long-Run Relationship Between Temperature and Sectoral Output

Next, cointegration was tested. This test was used to find out if there exists a long-run relationship between sectoral output and temperature. Rainfall is not taken into account in this test as it is a stationary variable at level. The p-values in Table 2 are the values in parenthesis and are highlighted by an asterisk at a 95% confidence level. Given that the p-values of the cointegration tests are less than 0.05 ($p < 0.05$), the null hypothesis of no cointegration is rejected. This result indicates that there is a long-run relationship between sectoral output and temperature, which answers the research objective of establishing if there is a long-run relationship between weather variability and sectoral output in Kenya. The long-run relationship between the temperature and sectoral output supports the findings of Nyangena (2016) who found that weather and output in Kenya had a long-run relationship.

## Effect of Weather Variability on Sectoral Output

After establishing the presence of a long-run relationship between sectoral output and temperature, a method known as fully modified ordinary least squares (OLS)

**Table 3** Fully modified ordinary least squares results

| Variable | Coefficient | P-value | Standard error |
|---|---|---|---|
| Temperature | −4.0229 | 0.7086 | 10.7312 |
| Rainfall | 0.1920 | 0.6976 | 0.4927 |
| CPIA | 0.7531 | 0.7624 | 2.4837 |
| Capital formation | −0.8404 | 0.4746 | 1.1710 |
| Labor | 8.2220[a] | 0.0102 | 3.1398 |

[a]Statistically significant at the 5% level of significance

was used to establish the relationship between weather variability and sectoral output. The results of the estimation are shown in Table 3.

The results in Table 3 show that there is a negative effect of temperature in degree Celsius on sectoral output, where a 1% increase in temperature leads to a 4.02% decrease in sectoral output holding all other factors constant. Regarding rainfall in millimeters, there is a positive relationship with sectoral output where a 1% increase in rainfall would result in a 0.19% increase in sectoral output holding all other factors constant. Colacito et al. (2019) also found an increase in temperature resulted in possible declines in economic growth. As seen in this chapter, there also exists an inverse relationship between temperature and sectoral output. Moreover, the inverse relationship between temperature and sectoral output is of a higher magnitude in comparison to the relationship between rainfall and sectoral output. This is consistent with the findings of Ochieng et al. (2016) who reported that temperature has a bigger effect on crop production in Kenya than rainfall. This chapter also supports the findings of Van-Passel et al. (2012) who argued that agricultural output is positively affected by rainfall when received on season and is good for overall economic performance. Even though this chapter found that rainfall has a positive effect on sectoral output, one should also keep in mind that continued heavy rain is likely to cause economic losses in Kenya as revealed by Ngecŭ and Mathu (1999) in their research. This research cannot particularly rule out this effect as it does not consider the seasonality of weather and the shocks therein.

The business regulatory environment as measured by the Country Policy and Institutional Assessment (CPIA) has a positive relationship with sectoral output where a 1% increase in the CPIA rating would result in an increase of sectoral output by 0.75% holding all other factors constant. The findings of this chapter also support the research carried out by Kimenyi et al. (2016) who reported that policy improvement would result in economic transformations and sustained economic growth in the sectors in Kenya.

Additionally, labor, which is the number of people employed, also has a positive relationship with sectoral output. A 1% increase in labor would result in an 8.22% increase in sectoral output holding all other factors constant. These results support the findings of Kimenyi et al. (2016) who found a positive correlation between human capital and productivity growth in Kenya. Further, they discuss the role of investment in education to improve the quality of human capital and spur rapid economic growth in the country.

Finally, capital formation and sectoral output have a negative relationship where a 1% increase in capital formation results in a 0.84% decrease in sectoral output holding all other factors constant. These results are different from the analysis carried out by Ongo et al. (2014) who find that technical progress and infrastructural development have a positive association with economic growth in the Economic and Monetary Community of Central Africa. Interestingly, the findings from Zeb (2013) indicate that the quality of capital falls as temperature increases and results in a drop in economic growth. With the exception of labor, all other variables in the model are not statistically significant. This indicates that weather variability in Kenya does not have a significant direct impact on sectoral output in the country. However, the annual nature of the data used in this research meant that the seasonal effects of weather variability on sectoral output were not captured, which may have led to the weather variables being statistically insignificant. Due to the unavailability of quarterly data (labor, gross fixed capital formation, and CPIA – business regulatory environment), this chapter was unable to take into account the seasonal effects of these variables.

## Conclusion

The results from this chapter indicate the presence of a long-run relationship between sectoral output in Kenya and temperature. Furthermore, there is an inverse relationship between temperature and sectoral output, which shows that an increase in temperature will lead to a decrease in sectoral output in Kenya. On the other hand, this research finds a positive relationship between rainfall and sectoral output, which reveals that an increase in rainfall in the country is likely to lead to an increase in GDP per sector. Kenya's economy is highly reliant on the agricultural sector with most small-scale farmers relying on rain-fed farming. As such, an increase in consistent rainfall will lead to increased agricultural output as long as the rainfall is not in excess. Additionally, findings from this chapter show that the business regulatory environment as measured by the World Bank CPIA rating has a positive relationship with sectoral output. This implies that a better policy and institutional framework will lead to higher output in the various sectors in Kenya. The number of people employed also has a positive relationship with sectoral output. Gross fixed capital formation, which measures the annual net increase in physical assets has a negative relationship with GDP per sector in Kenya.

With the evidence gathered from this research, it can be concluded that weather variability has an economic effect on output from the different sectors in Kenya. This suggests that the Kenyan government needs to take a keen interest in understanding and addressing the effects of weather variability and in a broader picture, the effect of climate change on each sector of economy. Some of the sectors may bear more of the brunt due to their dependent nature on weather. However, none of the sectors in the Kenyan economy will be left unaffected due to the effects of changing weather patterns and in the long term, climate change. Therefore, the Kenyan government should ensure that it fully supports the National Adaptation Plan, which is supported by the Climate Change Act that was enacted into law in May 2016. Additional support and increased partnerships will also be required for Kenya to achieve its climate adaptation goals. For example, Kenya needs to actively take a broader perspective that encompasses other countries including those outside the African continent in its environmental conservation discussions. This is because the impact of climate change affect all countries across the globe. As much as temperature and rainfall levels in Kenya are not at extreme levels, there has been a notable change in the weather patterns and the effects are starting to be felt. Given this, Kenya should consider being more aggressive in its reforestation campaign throughout the country. This will help to remove excess carbon dioxide emissions from the atmosphere, thus reducing the severity of weather pattern changes and in the long-run, mitigate climate change. Further research could be carried out to investigate the effect of weather variability on the Kenyan economy using data with a higher frequency. This would help to factor in the effect of climatic seasons in the country on the economy.

## References


Abidoye B, Odusola A (2015) Climate change and economic growth in Africa: an econometric analysis. J Afr Econ 25:1–25. https://doi.org/10.1093/jae/eju033

Awuor C, Orindi V, Adwera O (2008) Climate change and coastal cities: the case of Mombasa, Kenya. Environ Urban 20:231–242

Barro R (1999) Determinants of economic growth: implication of the global evidence for Chile. Cuad Econ 36(107):443–478

Colacito R, Hoffmann B, Phan T (2019) Temperature and growth: a panel analysis of the United States. JMCB 51:313–368. https://doi.org/10.1111/jmcb.12574

Dell M, Jones B, Olken B (2012) Temperature shocks and economic growth: evidence from the last half century. Am Econ J Macroecon 4:66–95

Deryugina T, Hsiang S (2014) Does the environment still matter? Daily temperature and income in the United States. NBER Working Paper (No. w20750; w20750)

Ferrara E (2015) The atlas of climate change: mapping the World's greatest challenge. University of California Press, Berkeley

Food and Agriculture Organization of the United Nations (2019) Kenya at a glance: the agriculture sector in Kenya. Retrieved from http://www.fao.org/kenya/fao-in-kenya/kenya-at-a-glance/en/

Hulme M (2009) Why we disagree about climate change. Cambridge University Press, Cambridge, UK

Idso C, Singer S (2009) Climate change reconsidered. The Heartland Institute, Chicago

Intergovernmental Panel on Climate Change (2014) Summary for policymakers. Contribution of working group II to the fifth assessment report of the intergovernmental panel on climate change. Cambridge University Press, New York

Kimenyi M, Mwega F, Ndung'u N (2016) Kenya economic growth, labor market dynamics, and prospects for a demographic dividend. In: Bhorat H, Tarp F (eds) Africa's lions: growth traps and opportunities for six African economies. Brookings Institution Press, Washington, DC, pp 4, 109–4, 144

Linke M, O'Loughlin J, McCabe T, Tir J, Witmer F (2015) Rainfall variability and violence in rural Kenya: investigating the effects of drought and the role of local institutions with survey data. Glob Environ Chang 34:35–47

Lorenz EN (1963) Deterministic nonperiodic flow. J Atmos Sci 20:130–141

Matsui T, Pielke RA (2006) Measurement based estimation of the spatial gradient of aerosol radiative forcing. Geospatial Res Lett 33:L11813

Montesquieu B, De Secondat C (1748) The spirit of laws. Batoche books: Kitchener, London

Mortimore M, Anderson S, Cotula L, Davies J, Faccer K, Hesse C, Wolfangel C (2009) Dryland opportunities: a new paradigm for people, ecosystems and development. International Union for Conservation of Nature and Natural Resources (IUCN), Gland

National Centers for Environmental Information (NCEI) (2018) http://www.ncei.noaa.gov/news/weather-vs-climate. Accessed 14 Mar 2020

Ngecũ WM, Mathu EM (1999) The El-Nino-triggered landslides and their socioeconomic impact on Kenya. Environ Geol 38(4):277–284

Nordhaus WD (2010) The economics of hurricanes and implications of global warming. Clim Change Econ 1:1–20

Nyangena OM (2016) The impact of weather variability on economic growth in Kenya. University of Nairobi Press. https://papers.ssrn.com/sol3/papers.cfm?abstract_id=3161257

Ochieng J, Kirimi L, Mathenge M (2016) Effects of climate variability and change on agricultural production: the case of small scale farmers in Kenya. NJAS - Wageningen J Life Sci 77:71–78

Okaka FO, Odhiambo BDO (2018) Relationship between flooding and out break of infectious diseases in Kenya: a review of the literature. J Environ Public Health. https://doi.org/10.1155/2018/5452938

Okoba B, Bryan E, Ringler C, Roncoli C, Silvestri S, Herrero MT (2013) Adapting agriculture to climate change in Kenya: household strategies and determinants. J Environ Manag 114:26–35

Ongo NB, Vukenkeng AW, Seppo MP (2014) Mechanisms of the influence of human capital on economic growth: a panel data analysis of CEMAC region. Asian Econ Financ Rev 4(5):625–640

Opere A (2013) Floods in Kenya. In: Paron P, Olago DO, Omuto CT (eds) Developments in earth surface processes. Elsevier, Oxford

Pedroni P (1996) Fully modified OLS for heterogeneous cointegrated panels and the case of purchasing power parity. Working paper in Economics 96(020). https://web.williams.edu/Economics/pedroni/WP-96-20.pdf

Smith A (1776) An inquiry into the nature and causes of the wealth of nations. (Cannan ed.) https://oll.libertyfund.org/titles/smith-an-inquiry-into-the-nature-andcauses-of-the-wealth-of-nations-cannan-ed-vol-1

Smith L (2002) What might we learn from climate forecasts? USA: Proc Natl Acad Sci 99(Suppl 1):2487–2492

Stephens G, Li JL, Wild M, Clayson C, Loeb N, Kato S, Andrews T (2012) An update on Earth's energy balance in light of the latest global observations. Nat Geosci 5(10):691–696

Sud Y, Walker G, Lau W (1999) Mechanisms regulating sea-surface temperatures and deep convection in the tropics. Geophys Res Lett 26(8):1019–1022

U.S. Global Change Research Program (2014) Global change. Retrieved from Extreme weather: https://nca2014.globalchange.gov/highlights/report-findings/extreme-weather

Van Passel S, Massetti E, Mendelson RO (2012) A ricardian analysis of the impact of climate change on european agriculture. Nota Di Lavoro, 83

World Bank (2012) Turn down the heat: why a 4C warmer world must be avoided. Potsdam Institute for Climate Impact Research and Climate Analytics. https://openknowledge.worldbank.org/handle/10986/11860

World Health Organization (2018, November 7) Climate change increases risk of outbreaks in Africa. Retrieved from WHO Africa: https://www.afro.who.int/news/climate-change-increases-risk-outbreaks-africa

Yanda PZ, Mubaya CP (2011) Managing a changing climate in Africa: local level vulnerabilities and adaptation experiences. African Books Collective. http://www.africanbookscollective.com/books/managing-a-changing-climate-in-africa

Zeb A (2013) Climate change and economic growth in Nordic countries: an application of smooth coefficient semi-parametric approach. Int J Soc Sci 11(3):159–171

Zivin J, Neidell M (2014) Temperature and the allocation of time: implications for climate change. J Labor Econ 32(1):1–26

# 8

# Assessment of Farmers' Indigenous Technology Adoptions for Climate Change Adaptation in Nigeria

Idowu Ologeh, Francis Adesina and Victor Sobanke

## Contents

Introduction .
Issue Description
Research Techniques and Findings ...
Devastating Effects of Climate Change on Smallholder Farmers
Indigenous Adaptation Techniques in Use in Nigeria
Constraints to the Development of Indigenous Adaptation Techniques in Nigeria
Indigenous Adaptation Techniques Contributing Factors .......
Conclusion
References

**Abstract**

Agriculture has shown a considerable capacity to adapt to climate change. Many adaptations occur autonomously without the need for conscious response by farmers and agricultural planners. However, it is likely that the rate and magnitude of climate change may exceed that of normal change in agriculture that specific technologies and management styles may need to be adopted to avoid the

I. Ologeh (✉)
Department of Environmental Management and Toxicology, Lead City University, Ibadan, Nigeria
e-mail: Ologeh.idowu@lcu.edu.ng

F. Adesina
Department of Geography, Obafemi Awolowo University, Ile-Ife, Nigeria

V. Sobanke
Research and Planning Department, National Centre for Technology Management, South West Office, Lagos, Nigeria

most serious of effects. Thus areas likely to be most vulnerable to climate variability can be spared from its impacts through implementation of appropriate adaptation measures such as development of indigenous technologies.

Six hundred farmers from the six geopolitical zones of Nigeria were surveyed and they all possess different indigenous adaptation strategies ranging from swamp farming (Oyo State), application of neem seed (Kaduna State), soil erosion control (Enugu State), rainwater harvesting (Taraba State), land improvement (Cross River State) to farmland management (Benue State). They all have simple but profound technologies driving these schemes with much success. These indigenous adaptation techniques are majorly constrained by inadequate financial resources. Indigenous technology adoption is affordable with high revenue potential.

**Keywords**

Indigenous technology · Climate change adaptation · Farmers · Nigeria

## Introduction

Historically, agriculture has shown a considerable capacity to adapt to changing climatic conditions. If climate change is gradual, the adaptation may go widely unobserved and the adjustment process largely independent (Parry et al. 2004). In the field of climate change, vulnerability has been described as the degree to which a system is susceptible to, or unable to cope with, adverse effects of climate change, including climate variability and extremes (Marrewijk 2011). Thus the need for climate change adaptation/mitigation measures to combat vulnerability. The major task of climate change adaptation and mitigation in agriculture is to produce more food efficiently and with clear reductions in greenhouse gas (GHG) releases from food manufacturing and marketing. "Adaptation" can be defined as societal or ecosystems efforts to prepare for or cope with future climate change. The coping methods can be defensive (i.e., being protective against adverse impacts of climate change), or adaptable (i.e., seizing the benefits of any advantageous effects of climate change) (USEPA 2014).

Nigeria along with other developing countries under UNFCCC is to focus on adaptation to make them able to cope with the new extremes in climatic regimes; they are also to concentrate on environmentally and economically sustainable development (Adesina and Odekunle 2011a). The argument is, the developing countries' growth is being impaired by climate change effect and these countries produce only a small fraction of the GHG that is causing climate change. They also have fragile adaptive capability because their economies are still growing, thus they are highly vulnerable to the impacts of climate change.

Climate change adaptation is not novel; history has told how human societies whether by migration, improved crop varieties, or diversifying housing types have repeatedly proven strong capacities for adapting to different climates and environmental changes (Adger et al. 2007). However, the current rate of global climate change is unusually high compared to past changes that society has experienced and thus new innovations are necessary to combat the challenge. In this era that the world is getting increasingly interdependent, adverse effects of climate change on one community or economic sector can have aftermaths around the world (United States Global Change Research Programme – USGCRP 2009).

Adaptation is globally important because climate change will not become history soon. Many greenhouse gases linger in the atmosphere for 100 years or more after their emission and because of their long-lasting effects, they will continue to warm Earth in the twenty-first century, even if additional greenhouse gases emission were to stop today. Therefore, steps must be taken now to prepare for, and respond to, the impacts of climate change that are already occurring, and those that are projected to occur in the years ahead (USCGRP 2009).

There is need for continuous actions to mitigate climate because there are limits to the ability to adapt. Adaptation alone on long-term basis may not be sufficient to cope with all the projected impacts of climate change, thus it will need to be continuously coupled with actions to lower greenhouse gas emissions (IPCC 2007).

Adaptation is based on the level of socioeconomic development of a country; the resilience and adaptive capacity of a country is dependent on its level of development with respect to economy and political stability. To improve adaptation strategies, a clear understanding of local susceptibilities to climate change and critical thresholds must be established. Adaptation strategies should be adequately flexible in order to accommodate future possible changes in climatic parameters which may be responsible for quick review of plans; the plan should be reviewed periodically.

A myriad of possible adaptation strategies for agriculture are available; some of the most prominent in Nigeria are innovative indigenous technology options ranging from improved crop varieties, composting of organic waste, recycling and waste minimization, improved cultivation techniques, and cover cropping (Eze and Osahon 2015).

Reactive adaptations are those which occur after the impacts of climate change have been experienced, while anticipatory adaptations are proactive, undertaken before the impacts are fully felt. Planned adaptations are generally anticipatory, but can also be reactive (i.e., climate change effects are experienced). Farmers have access to various adaptations depending on their local environments and the specific farming system.

Some impact studies have suggested "adaptations" involving reductions and expansions of agricultural zones. This implies that some farmers would relocate, or others would completely change their type of farming while others would stop operations in some locations. Also, in other locations, there would be new farmers and some existing landowners would try new types of farming (Adesina and Odekunle 2011b). Although adaptations can be strategic at the farm level, the term "planned adaptation" is commonly used to describe actions taken by governments as a conscious policy response. Probable governments' planned adaptations include reinforcement of technological adaptations, such as crop development and early warning systems, promotion of land and water use options, changes in diversification or intensity of production aid, transformed financial support in established programs, and impromptu compensation.

## Issue Description

Smallholder farmers were facing a lot of climate change-induced challenges on their farms of which flooding, pest and disease invasion, high temperature, and low crop yield are the most prominent. The Nigerian government has promoted various adaptation measures as its effort to curtail these challenges. These include provision of improved crop varieties, fertilizers, irrigation schemes, and geodata (Adefolalu 2007). Wisner et al. (2004) report that farmers' vulnerability is not determined by the nature and magnitude of climate change as such, but by the interplay of the societal capacity to adapt and/or recuperate from environmental change. The adaptation capacity and degree of exposure is connected to environmental changes and also to changes in societal aspects such as land use and cultural practices. Most studies on climate change and agriculture in Africa have concentrated on actual and projected impacts as well as farmers' coping/adaptation strategies (Adejuwon 2004). There has been little or no work in the area of local/indigenous adaptation technologies and their challenges or success. This chapter will therefore attempt, through a questionnaire survey, Focus Group Discussions, and review of relevant literature, to fill this gap.

The objective of this chapter is to assess indigenous technology adaptation options being used by smallholder farmers in Nigeria.

## Research Techniques and Findings

The survey area is Nigeria; for ease of data collection, the major food-producing state of each geopolitical zone in the country is sampled for the survey. After relevant literature and National Bureau of Statistics consultation, the following six states were selected for the survey:

North-Central Zone – Benue State
North East Zone – Kaduna State
North-West Zone – Taraba State
South-East Zone – Enugu State
South-South Zone – Cross River
South-West Zone – Oyo State

It is very important to assess the various indigenous technology adaptations in use by smallholder farmers in Nigeria so as to know which can be fully developed for national adoption. To achieve this objective, 600 questionnaires were administered to smallholder farmers in the six states to survey their various indigenously designed adaptation options to climate change. The questionnaires were designed to collect information on farmers' biometric data, years of experience and skills, farm size, types of crops cultivated, educational level, experiences on climate change effects, and indigenous adaptation techniques. Secondary data was obtained from NBS and the Agricultural Development Agencies of each of the States. The retrieved questionnaires were imputed and analyzed using Microsoft Excel spreadsheet 2016 and the results are presented in frequencies and percentages.

A total of six hundred questionnaires were administered to smallholder farmers in the six states, and all were retrieved through the assistance of agricultural extension workers who administered them, the results are presented in Table 1.

In order to obtain realistic data, matured farmers are targeted, the average age is 53 years old. Majority (86%) of the farmers have more than 30 years of experience in farming with 60% of them owning more than one hectares of land. Female farm owners (42%) are mostly found in Southern Nigeria while men majorly own farmland in the north, and employ numerous women laborers. The most grown crops across Nigeria as detected in the data are cassava, maize, yam, rice, and vegetables.

Almost all the farmers (97%) complained of the various losses they have incurred due to devastation effects of climate change ranging from drought encroachment in the north to flooding in the south. The totality of the respondent farmers have poor knowledge of the scientific explanations behind climate change, but experience taught them that nature (weather) is no longer their friend and they have to devise strategies to make it work in their favor. Their understanding is that nature (or the gods) is fighting them through the massive attack of pests and diseases, flooding, erratic rainfall, high temperature, and low yield. As a result, they devise different techniques to adapt to climate change effects including sacrificing to the gods. These indigenous adaptation techniques vary from zone to zone, depending on the

**Table 1** Socioeconomic characteristics of sampled smallholder farmers

| Variables | Categories | Benue | Cross River | Enugu | Kaduna | Oyo | Taraba |
|---|---|---|---|---|---|---|---|
| | | Freq/% | Freq/% | Freq/% | Freq/% | Freq/% | Freq/% |
| Sex | Male | 67 | 41 | 46 | 72 | 52 | 66 |
| | Female | 33 | 59 | 54 | 28 | 48 | 34 |
| Age(years) | ≤ 40 | 2 | 3 | 0 | 1 | 0 | 4 |
| | 41–50 | 39 | 37 | 30 | 26 | 29 | 29 |
| | 51–60 | 54 | 51 | 59 | 58 | 61 | 59 |
| | 61–70 | 5 | 9 | 11 | 15 | 10 | 8 |
| Marital | Single | 3 | 2 | 4 | 0 | 7 | 0 |
| Status | Married | 97 | 96 | 91 | 99 | 89 | 98 |
| | Widowed | 0 | 2 | 5 | 1 | 4 | 2 |
| Levels of | Informal | 61 | 14 | 17 | 48 | 12 | 68 |
| Educational | Primary | 32 | 52 | 32 | 38 | 23 | 30 |
| Attainment | Secondary | 07 | 34 | 51 | 14 | 65 | 02 |
| Farm | ≤ 1 | 25 | 63 | 67 | 31 | 33 | 21 |
| Size(ha) | 2–5 | 57 | 26 | 29 | 42 | 55 | 56 |
| | Above 5 | 18 | 11 | 4 | 27 | 12 | 23 |
| Farming | ≤ 30 | 13 | 19 | 18 | 11 | 17 | 6 |
| Experience | 31–40 | 60 | 55 | 53 | 58 | 56 | 62 |
| (years) | Above 41 | 27 | 26 | 29 | 31 | 27 | 32 |

Source: Fieldwork

prevalent climate change effect in the area. Swamp farming was devised in Oyo State, discovery, and application of neem seed to enhance soil fertility in Kaduna State, improved soil erosion control in Enugu State, rainwater harvesting in Taraba State, intentional cover cropping in Cross River State to farmland management (organic manure) in Benue State. They all have simple but profound technologies driving these schemes which will be discussed in details.

This finding harmonized with that of Eze and Osahon (2015) who listed adaptation strategies as improved crop varieties, composting of organic waste, recycling, and waste minimization, improved cultivation techniques, and cover cropping.

## Devastating Effects of Climate Change on Smallholder Farmers

All the smallholder farmers surveyed have in their more than 30 years of experience suffered different adverse effects of climate change. The prominent effect in Northern Nigeria is erratic rainfall and unusual heat (heatwave). They are also faced with desert encroachment advancing into Nigeria from Niger republic and pest invasion. Their counterparts in the South are majorly faced with flooding, infertility, increase in temperature, and pest invasion. The major effect in North Central is the increase in temperature, soil infertility, and pest invasion.

The finding is in agreement with the work of Lybbert and Sumner (2010) that temperature increase is an indication of climate change, and also the work of Adegoke et al. (2014), who stated that weeds, pests, and fungi thrive under warmer temperatures, wetter climates, and increased carbon dioxide ($CO_2$) levels.

The different climate change effects being experienced in different zones in Nigeria dictate the corresponding indigenous adaptation techniques, and these are presented in Tables 2 and 3.

## Indigenous Adaptation Techniques in Use in Nigeria

### North-Central Zone – Benue State

Benue State is the food basket of Nigeria, the state is popular for large-scale food and fruit production including yam, cassava, sweet potato, beans, maize, millet, guinea corn, vegetables, soybeans, rice, and citrus. The smallholder farmers in Benue State are faced with climate change impacts such as soil infertility, increase in temperature, pests and diseases invasion, crop failure, and increased weed. To mitigate/adapt to these effects, the adaptation strategies common to smallholder farmers in the state are mixed cropping, growing pest/disease-resistant crop varieties, use of cover crops, and making bigger ridges.

The prominent indigenous adaptation practice in the state is local farmland management. The Benue farmers devised a farmland management practice to adapt to climate change. The practice entails conscious efforts to reduce temperature on farmlands and is very similar to organic agriculture. It is a complete management system with high organic matter content (mulching), soil cover (cover crops/tree

**Table 2** Devastating effects of climate change on smallholder farmers

| Variables | Benue | Cross river | Enugu | Kaduna | Oyo | Taraba |
|---|---|---|---|---|---|---|
| Mean max 5 | Mean (M) | Mean (M) | Mean (M) | Mean (M) | Mean (M) | Mean (M) |
| Soil infertility | 4.2 | 2.1 | 1.9 | 4.4 | 2.7 | 3.0 |
| Increase in temperature | 4.3 | 4.0 | 4.3 | 4.2 | 4.4 | 4.1 |
| Pest invasion | 4.6 | 4.7 | 4.5 | 4.4 | 4.5 | 4.4 |
| Crop failure | 3.6 | 4.2 | 3.9 | 4.0 | 3.5 | 3.7 |
| Increased weed | 3.9 | 3.8 | 3.1 | 2.8 | 3.4 | 4.2 |
| Drought | 1.2 | 0.8 | 1.5 | 3.8 | 1.1 | 3.7 |
| Land degradation | 2.3 | 2.8 | 2.5 | 3.6 | 3.1 | 3.2 |
| Flooding | 3.2 | 3.8 | 4.1 | 3.6 | 3.9 | 2.9 |
| Access to water | 2.4 | 0.5 | 1.5 | 3.7 | 1.2 | 3.4 |
| Decrease in soil moisture | 3.1 | 1.5 | 2.4 | 3.2 | 2.1 | 3.7 |
| Erosion | 3.4 | 3.2 | 3.9 | 3.1 | 3.8 | 2.9 |

Source: Fieldwork

planting), and high soil fertility (crop rotation, organic manures, and legume planting) thus retaining nutrient and water, building more floods, drought, and land degradation resilient soils.

## North West Zone – Kaduna State

Kaduna is a state in which 80% of the people are actively engaged in farming. They produce crops ranging from yam, maize, beans, guinea corn, millet, rice, and cassava. The prevalent climate change impacts in the state are drought, pests and diseases invasion, crop failure, land degradation, increase in temperature, flooding, access to water and soil infertility. The widely adopted adaptation practices are mixed farming, mixed cropping, growing drought-resistant crop varieties, growing pest-resistant crop varieties, crop rotation, irrigation, roof water harvesting, and making bigger ridges.

The unique indigenous adaptation method in use in the State is the application of neem seed for pest control (used as fumigant, pesticide), compost (used as fertilizer, manure), and soil fertility (used as soil conditioner and urea coating agent).

## North-East Zone – Taraba State

Taraba State just as his North-West counterpart is 80% agrarian. They produce crops ranging from maize, millet, sorghum, rice, yam, cassava, and sweet potatoes. They experience climate change effects such as high rate of weeds, drought, decrease in soil moisture, increase in temperature, decrease in crop yields, and high rate of pests and disease incidence. The adaptation measures generally in use in the State are growing drought-resistant crop varieties, crop rotation, irrigation, integration of livestock farming system, changing crop varieties, mulching, and intercropping.

**Table 3** General adaptation techniques in use in Nigeria

| Variables | Benue | Cross river | Enugu | Kaduna | Oyo | Taraba |
|---|---|---|---|---|---|---|
| Mean max 5 | Mean (M) | Mean (M) | Mean (M) | Mean (M) | Mean (M) | Mean (M) |
| Mixed cropping | 3.7 | | | 3.8 | | |
| Growing pest/disease resistant crop varieties | 3.9 | 3.7 | 3.7 | 4.1 | 4.4 | 3.4 |
| Use of cover crops | 3.6 | 3.7 | 2.9 | 3.3 | 3.4 | 3.1 |
| Making bigger ridges | 3.5 | 3.3 | 3.3 | 3.7 | 3.4 | 3.3 |
| Mixed farming | 3.2 | 2.7 | 2.8 | 4.4 | 3.1 | 3.4 |
| Growing drought resistant crop varieties | 3.1 | 2.1 | 2.8 | 3.6 | 2.8 | 3.5 |
| Mulching | 3.1 | 4.2 | 3.3 | 3.2 | 3.9 | 4.4 |
| Crop rotation | 3.4 | 2.1 | 2.4 | 3.7 | 2.7 | 4.1 |
| Irrigation | 3.4 | 3.3 | 3.4 | 4.0 | 4.2 | 4.1 |
| Roof water harvesting | 3.1 | 2.7 | 2.8 | 3.7 | 2.5 | 3.1 |
| Integration of livestock farming system | 2.3 | 1.2 | 1.8 | 3.3 | 2.9 | 3.6 |
| Changing crop varieties | 2.6 | 2.2 | 2.4 | 3.1 | 3.3 | 3.6 |
| Intercropping | 3.2 | 2.3 | 2.8 | 3.1 | 3.3 | 3.8 |
| Use of pesticides/herbicides | 2.4 | 2.5 | 3.9 | 2.7 | 4.2 | 3.1 |
| Construction of drainage systems | 1.2 | 3.7 | 4.2 | 2.2 | 2.9 | 2.8 |
| Contour cropping | 2.1 | 2.0 | 3.9 | 3.1 | 3.4 | 3.1 |
| Diversification in crop planting | 2.4 | 3.6 | 3.8 | 2.8 | 3.2 | 3.0 |
| Ridge construction | 2.8 | 2.7 | 3.3 | 3.4 | 3.3 | 3.5 |
| Crop substitution | 3.4 | 2.1 | 2.8 | 3.3 | 3.4 | 3.7 |
| Changing planting dates | 3.1 | 3.1 | 3.0 | 3.3 | 3.2 | 3.5 |

Source: Fieldwork

The indigenous adaptation practice unique to Taraba State is rain harvesting. Access to water is an issue in the state and thus the need for irrigation. Rain harvesting is used to augment the water used for irrigation. The most common is roof water harvesting which is channeled into catchment tanks or concrete reservoirs.

## South-East Zone – Enugu State

Enugu State has a diversified economy dominated by agriculture. Major crops produced in the State are yam, cassava, maize, rice, cowpea, sweet potatoes, and plantain. Their crop production is impaired by climate change effects such as increase in temperature, flood, soil erosion, pests and diseases invasion, and crop failure. Adaptation practices widely in use in the State include planting pest/disease-resistant crop varieties, use of pesticides/herbicides, construction of drainage systems, contour cropping, and diversification in crop planting.

Soil erosion is the major problem affecting smallholder farmers in Enugu State, and its volume was aggravated by the effects of climate change. Their key indigenous adaptation strategy is soil erosion control using stone and sandbags to divert erosion away from the farms, ridging, and use of wrapped weeds or grasses to cover planted heaps. Stems from previous harvests are also arranged or scattered on the soil in bands to reduce soil erosion. Tree planting on the borders of the farm also prevents soil erosion as the trees shield the farm from direct impact of rainfall.

### South-South Zone – Cross River

Cross River State famously known for tourism and fishing is also fully involved in agriculture. The crops mainly cultivated in the State are cassava, yam, plantain, rice, and maize. Although it is a coastal state, they also have their share of climate change effects. They are affected by increased weeds, pests and diseases invasion, increase in temperature, crop failure, and flooding. They adapt by planting pest-resistant crop varieties, diversification of crops, planting cover crops, mulching, and construction of drainage systems.

Indigenous land improvement methods are the key indigenous adaptation practice in Cross River State. These methods include organic addition to the land (mulching, compost, and manure), cover tree planting, planting of legume crops (mixed cropping), and polyculture. Water harvesting from runoffs also helps to maintain soil moisture during dry season.

### South-West Zone – Oyo State

Oyo State is an industrialized state with many educational institutions. It also enjoys diversified economy with agriculture as the major occupation. Smallholder farmers in Oyo State are faced with climate change effects like their counterparts in other states, this includes flooding, pests and diseases invasion, soil erosion, crop failure, and increased temperature. Their general adaptation methods are ridge construction, use of pesticides and pest-resistant crop variety, irrigation, crop substitution, and changing planting dates.

The major captivating indigenous adaptation method practiced by the smallholder farmers in Oyo State is swamp farming. There are lots of swamp areas in Oyo State, where smallholders resort to when there is prolonged delay in rainfall season. This initiative allows early and late farming as there is constant access to water. Some farmers in Oke Ogun area of the state channel their domestic wastewater to wet their vegetable farms all year round.

These findings are in agreement with that of Adger et al. (2007) that listed adaptation strategies as migration, improved crop varieties, or building different types of shelter, while Anríquez and Stamoulis (2007) list include changing the timing of operations, adoption of conservation tillage practices, and diversification in production systems, improvement of irrigation schemes, modification of farm support programs, and development of new plant varieties.

## Constraints to the Development of Indigenous Adaptation Techniques in Nigeria

The various indigenous adaptation strategies assessed have developmental constraints. The major is inadequate financial resources; adaptation practices are cost-intensive and these smallholder farmers cannot afford elaborate adaptation strategies. Even the indigenous methods need funding to develop, e.g., the planting and processing of neem seed in Kaduna State and soil erosion control in Enugu State. Other constraints include inadequate farm labor due to express increase in rural-urban migration; poor extension services, insufficient drought-resistant varieties, and strict adherence to local varieties (see Table 4).

## Indigenous Adaptation Techniques Contributing Factors

The results of the partial least squares (Ringle et al. 2005) are presented in Fig. 1, and showed the factor loadings for all observed variables, $R^2$ value of the unobserved endogenous (dependent) variable as well as regression coefficients between exogenous and endogenous unobserved variables. Nine (9) observed variables

**Table 4** Constraints to the development of indigenous adaptation techniques in Nigeria

| Variables | Benue | Cross river | Enugu | Kaduna | Oyo | Taraba |
|---|---|---|---|---|---|---|
| | Freq /% | Freq /% | Freq /% | Freq /% | Freq /% | Freq /% |
| Finance | 93 | 92 | 88 | 95 | 91 | 86 |
| Farm labor | 34 | 62 | 57 | 37 | 48 | 33 |
| Poor extension services | 23 | 42 | 38 | 31 | 54 | 48 |
| Insufficient drought-resistant varieties | 76 | 84 | 63 | 67 | 66 | 52 |
| Strict adherence to local varieties | 62 | 33 | 28 | 56 | 14 | 51 |

Source: Fieldwork

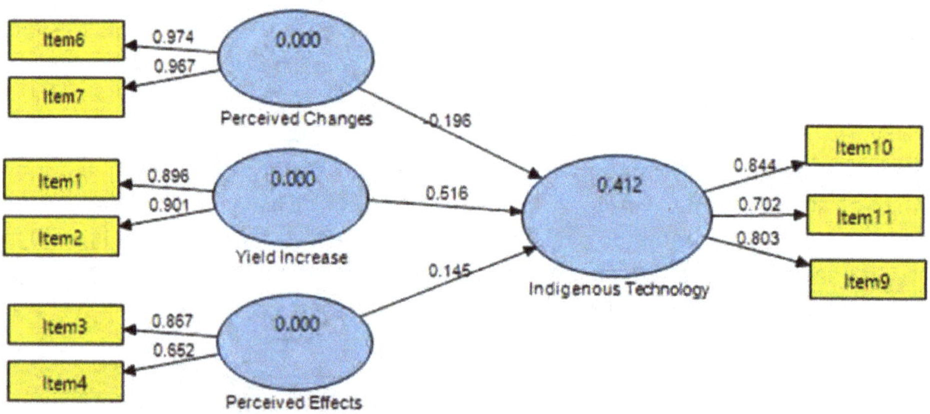


**Fig. 1** Structural model of indigenous technology adoption

with high loadings were retained for further analysis, while items with low factor loadings were removed. The $R^2$ value of 0.412 indicates that about 41% variance observed in the choice of indigenous adaptation techniques employed by farmers can be explained by farmers' perceived effect of climate change, need for crop yield increase, and perception of changing weather.

Reliability and validity of unobserved variables is presented in Table 5. The results show that composite reliability (CR) which indicates the convergent validity in all the constructs is adequate and above 0.7 the minimum threshold (Hair et al. 2010). The average variance extracted (AVE) which is a more conservative method of measuring convergent validity than CR (Malhotra and Dash 2011) is also within the recommended threshold (>0.5). On the contrary, only the value of Cronbach's alpha for perceived effects was below the recommended threshold (0.7).

The results of the T-statistics showing the significant level of regression coefficients are presented in Table 6. This result showed that indigenous adaptation techniques by respondents is influenced by their perceived changes (2.703; $p < 0.05$) and yield increase (6.665; $p < 0.05$). The result suggest that respondent's choice of indigenous adaptation techniques depends on the aspect of perceived changes in climatic parameters as well as the usefulness of climatic information.

**Table 5** Reliability and validity of unobserved variables

| | AVE | Composite reliability | Cronbach's alpha | Communality | Redundancy |
|---|---|---|---|---|---|
| Indigenous technology | 0.617 | 0.828 | 0.700 | 0.617 | 0.056 |
| Perceived changes | 0.942 | 0.970 | 0.938 | 0.942 | |
| Perceived effects | 0.589 | 0.737 | 0.317 | 0.589 | |
| Yield increase | 0.807 | 0.893 | 0.761 | 0.807 | |

**Table 6** Path analysis

| | Original sample (O) | Sample mean (M) | Standard deviation (STDEV) | Standard error (STERR) | T-statistics (\|O/STERR\|) |
|---|---|---|---|---|---|
| Perceived changes -> indigenous technology | −0.196 | −0.198 | 0.072 | 0.072 | 2.703 |
| Perceived effects -> indigenous technology | 0.145 | 0.156 | 0.081 | 0.081 | 1.782 |
| Yield increase -> indigenous technology | 0.516 | 0.519 | 0.077 | 0.077 | 6.665 |

## Conclusion

The chapter examined the various climate change effects affecting smallholder farmers across Nigeria. The general adaptation techniques adopted in each geopolitical zone of the nation were assessed as well as peculiar indigenous adaptation techniques initiated and in use in each zone. The constraints to the development of the indigenous adaptation techniques were also investigated. Going forward, these indigenous techniques need to be developed and commercialized; the state and federal government agricultural schemes and agencies can fund this project and support the efforts of the smallholder farmers. It is also essential for a functional link to be established between the farmers' indigenous innovation and the academia to foster research and development.

## References

Adefolalu DO (2007) Climate change and economic sustainability in Nigeria. In: Paper presented at the international conference on climate change and economic sustainability held at Nnamdi Azikiwe University, Awka, Nigeria. 12–14 June 2007

Adegoke J, Araba A, Ibe C (2014) National agricultural resilience framework. Federal ministry of agriculture and rural development, Abuja

Adejuwon SA (2004) Impacts of climate variability and climate change on crop yield in Nigeria. In: Paper presented at the stakeholder's workshop on assessment of impacts and adaptation to climate change, conference centre, Obafemi Awolowo University, Ile-Ife 20–21 September 2004

Adesina FA, Odekunle TO (2011a) Climate change and adaptation in Nigeria: some background to Nigeria's response – part I. 2011. International conference on environmental and agriculture engineering IPCBEE vol.15 (2011). IACSIT Press, Singapore

Adesina FA, Odekunle TO (2011b) Climate change and adaptation in Nigeria: some background to nigeria's response – part III. 2011. International conference on environmental and agriculture engineering IPCBEE vol.15 (2011). IACSIT Press, Singapore

Adger WN, Agrawala S, Mirza MMQ, Conde C, O'Brien K, Pulhin J, Pulwarty R, Smit B, Takahashi K (2007) Assessment of adaptation practices, options, constraints and capacity in climate change 2007: impacts, adaptation, and vulnerability. Contribution of working group II to the fourth assessment report of the intergovernmental panel on climate change. Eze and Osahon; AJAEES 6(1):45–55, 2015; Article no.AJAEES.2015.061 Asian Journal of Agricultural Extension, Economics & Sociology 6(1):45–55, 2015; Article no.AJAEES.2015.061 ISSN: 2320–7027

Anríquez G, Stamoulis S (2007) Rural development and poverty reduction: is agriculture still the key? Journal of Agricultural and Development Economics 4(1):5–46

Eze S and Osahon E (2015) Indigenous mitigation and adaptation to climate change among small holder farmers in arochukwu area of Abia State, Nigeria. Asian Journal of Agricultural Extension, Economics & Sociology 6(1): 45–55. Article no. AJAEES.2015.061 ISSN: 2320-7027

Feenstra JF, Burton I, Smith JB, Tol RS (eds.) (1998) Handbook on Methods for Climate Change Impact Assessment and Adaptation Strategies. United Nations Environment Programme; Vrije Universiteit, Amsterdam. 8–1 – 8–39

Hair JF, Black WC, Babin BJ, Anderson RE (2010) Multivariate data analysis: international version. Pearson, New Jersey

Intergovernmental Panel on Climate Change- IPCC (2007) Summary for policymakers, in climate change 2007: impacts, adaptation and vulnerability. Contribution of working group II to the fourth assessment report of the intergovernmental panel on climate change. Cambridge University Press, Cambridge, UK, p 17

Lybbert T, Sumner D (2010) Agricultural technologies for climate change mitigation and adaptation in developing countries: policy options for innovation and technology diffusion. International Centre for Trade and Sustainable Development (ICTSD), Geneva

Malhotra NK, Dash S (2011) Marketing research an applied orientation. Pearson Publishing, London

Marrewijk L (2011) Climate smart agriculture in the Mutale Basin, South Africa. Sustain Livelihoods Biodivers Dev Countries 4(2):5–6

Parry ML, Rosenzweig Cynthia, Iglesias Ana, Livermore M, Fischer Günther (2004) Effects of climate change on global food production under SRES emission and socio-economic scenarios. Global Environmental Change 14:53–67. https://doi.org/10.1016/j.gloenvcha.2003.10.008

Ringle CM, Wende S, Will A (2005) SmartPLS 2.0.M3. SmartPLS, Hamburg, https://www.smartpls.com/smartpls2

United States Environmental Protection Agency -USEPA (2014) Climate change impacts and adapting to change. Retrieved March 31, 2015, from http://www.epa.gov/climatechange/impacts-adaptation/

United States Global Change Research Program -USGCRP (2009) Global climate change impacts in the United States. In: Karl TR, Melillo JM, Peterson TC (eds) United States global change research program. Cambridge University Press, New York

Wisner B, Blaikie P, Cannon T, Davis I (2004) At risk: natural hazards; people's vulnerability and disasters. 2nd Edition. London: Routledge

# 9

# Climate Change, Rural Livelihoods and Ecosystem Nexus: Forest Communities in Agroecological Zones of Nigeria

Olushola Fadairo, Samuel Olajuyigbe, Tolulope Osayomi, Olufolake Adelakun, Olanrewaju Olaniyan, Siji Olutegbe and Oluwaseun Adeleke

## Contents

Introduction
- Forest Livelihoods and the Challenge of Changing Climate
- Agriculture, Climate Change, and Food Security
- Climate Change and Drivers of Social Vulnerability in Nigeria

Theoretical Framework
- Sustainable Livelihood Framework

Chapter Approach

Effects of Climate Variability on Cropping Calendar in Mangrove, Rainforest, and Savannah Agroecological Zones of Nigeria
- Perceived Trend in Agriculture, Forest Area, and Population as Primary Drivers of Social Vulnerability
- Perceived Effects of Drivers of Climate Change on Household Livelihood

O. Fadairo (✉) · O. Adelakun · S. Olutegbe · O. Adeleke
Department of Agricultural Extension and Rural Development, University of Ibadan, Ibadan, Nigeria
e-mail: dairom2@gmail.com; flakyonline@yahoo.com; siji004u@yahoo.com; aladeoluwaseun@yahoo.com

S. Olajuyigbe
Department of Forest Production and Products, University of Ibadan, Ibadan, Nigeria
e-mail: lekito2001@yahoo.com

T. Osayomi
Department of Geography, University of Ibadan, Ibadan, Nigeria
e-mail: osayomi@yahoo.com

O. Olaniyan
Department of Economics, University of Ibadan, Ibadan, Nigeria
e-mail: lanreolaniyan@yahoo.co.uk

Causes and Consequences of Environmental Degradation in the Nigeria Agroecological Zones
Farming Communities' Needs for an Enduring Adaptation to Climate Change in Nigeria
Conclusion and Recommendation
References

**Abstract**

A top-bottom approach where local problems are treated in isolation has proven ineffective in achieving sustainable development. The need for inclusive approaches to managing the demand for arable lands, forest resources, and the problems of resource exploitation and climate change calls for local understanding of these elements' interrelationship. Understanding the interrelationships among climate change, agriculture, and the ecosystems in different agroecological zones in Nigeria was the purpose of this chapter. Deforestation and forest degradation analysis approach was utilized. One state and two forest communities from each of the rainforest, savannah, and mangrove agroecological zones were purposively focused in this chapter based on forest distribution and cover. Focus group discussions involving 252 male and female farmers using 30 years as reference were used to garner relevant information. Climate variation caused a slight modification in cropping schedules of farmers due to prolonged dry season, mainly in the savannah region. Farmers engaged in mixed farming and also cultivate more hardy crops like cassava in response to climate uncertainties. Especially in the mangrove and savannah, ecosystem components such as agriculture and population showed increasing trends over the years as forest cover reduces. Downward trend in charcoal production was limited to mangrove and rainforest zones as fishing and hunting becomes vulnerable livelihoods across the zones. The degree and progression of climate change effects on the ecosystem in Nigeria agroecological zones is largely comparable and have both desirable and adverse livelihood outcomes. Affordable insurance policy, credit, agri-inputs, favorable forest regulatory framework, and youth empowerment supports would enhance sustainable adjustment to climate change.

**Keywords**

Forest communities · Cropping calendar · Agroecology · Climate change · Vulnerability · Rural livelihoods · Nigeria

## Introduction

Nigeria is seriously threatened by climate change with a significant proportion of its terrestrial ecosystem on dry land mass which is frequently affected by desertification, sheet erosion, and droughts. The coastal and mangrove agroecological zones in the south are also prone to incessant flooding because of their proximity to the Atlantic Ocean, riverine nature of the setting, the very low altitude, and all-year-round and high volume of rainfall. In recent times, variations in climatic conditions have resulted in undesirable effects on food production and nutritional security. Unfortunately, the country has very weak adaptive strategies and capacity to mitigate the effects of a changing climate. Presently, the impacts of rising temperatures and rainfall variability on farming are being felt across major agroecological zones in Nigeria (Ayanlade et al. 2018).

Agricultural systems are dependent on ecosystem services such as nutrient cycling, pollination, soil fertility, hydrological balances, and biological pest control which ensure a balance in the ecosystem (Power 2010). However, agricultural intensification in the last century has distorted the ecosystem equilibrium and led to loss of ecological integrity, land degradation, and loss of environmental services provided by the ecosystems. These conditions are further worsened due to increasing effects of climate instability (Pretty et al. 2011). For instance, environmental problems such as groundwater depletion, variability in the onset and amounts of rainfall, increase in concentration of greenhouse gases, soil degradation, depletion of pollinators' habitat, which all have negative consequences on sustainable agriculture, are climate change-induced.

Forest communities, which are highly vulnerable to these adverse effects, are occupied by low-income earners who depend on the ecosystem for their income, food, nutritional security, and livelihoods. Hence, these rural populations will be seriously affected by climate change, with little or no resources to adapt or mitigate its effects. It has been reported that the livelihoods of these communities are made vulnerable by land use variation such as continuous grazing and monoculture plantation. For example, in coastal and mangrove regions, there is a shortage of food resources obtained from streams coupled and agricultural instability due to increased flooding (Ward et al. 2016). Similarly, savannah and rainforest agroecosystems are recording a decline in agricultural production outputs (Ayanlade et al. 2018).

Natural resource utilization forms the base of most livelihoods in developing countries including Nigeria. However, forest resources are gradually being depleted due to the pressures of degradation and deforestation, poverty, urbanization, and poor management (Azeez et al. 2010; Saka et al. 2013). In Nigeria and most parts of Africa, shifting cultivation among small-holder farmers results in large-scale habitat destruction (Cooper et al. 2008). For instance, over 75% of the Nigerian population still lives in rural areas with vast areas of forest vegetation and depend on extensive rain-fed farming as well as short fallows for their sustenance. However, this dependence is limited by loss of forest biodiversity, climate change, and exposure of fragile soils (Azeez et al. 2010).

Frequent changes in climate parameters affect the livelihoods of rural populations and poses challenges to food security, survival, and economic development (Tompkins and Adger 2004). For instance, savannah and tropical forest zones have experienced a dramatic decrease in annual rainfall and an increase in the length of dry season and rainfall variability (Malhi and Wright 2004; Veenendaal and Swaine 1998). It is therefore pertinent to investigate the forest-dependent communities' responses to the hazards posed by climate uncertainties on their environment and livelihoods (Lindner and Pretzsch 2013), as this will have important implications for sustainable development in the near future.

The capability of farming communities and agricultural stakeholders to manage the challenges and prospects of present climatic patterns must primarily be improved in order to enhance their adaptive capacity and reduce their exposure to the undesired effects of changing climatic conditions (Cooper et al. 2008; Tompkins and Adger 2004). The challenges are complex and call for integrative learning-oriented approaches that emerge from the bottom-up that will enable successful mitigation and adaptation. These approaches offer pathways for vulnerable communities to engage in developing response policies and ensure that there is room for change in those policies (Lindner and Pretzsch 2013).

Adaptive management is a cyclic, multiple stakeholder learning-oriented approach to the management of complex environmental problems such as climate change. The introduction of such approaches would encourage multi-stakeholder participation and the integration of all sectors in decision-making, policy formulation, and implementation (Stringer et al. 2006). Unfortunately, most policy implementers adopt a top-bottom approach where local problems are treated in isolation, and this method has proven to be ineffective in yielding or sustaining solutions. Therefore, the need for all-encompassing approaches to manage the demand for arable lands, forest resources, and the problems arising from resource exploitation and climate change (Lindner and Pretzsch 2013; Tompkins and Adger 2004). A step towards achieving this is to understand that the poor and vulnerable themselves are key actors in identifying important areas of their own livelihoods and solutions to their challenges.

For instance, rural communities have in times past developed indigenous technologies which have assisted in mitigating the risks associated with climate variability. These technologies are represented in local customs, traditions, and heritages, constituting a testimony of how societies have thrived well in various environments (Azeez et al. 2010; Vidaurre de la Riva et al. 2013). Therefore, research needs to integrate climate change impacts with sustainable agriculture in a stressed ecosystem. Such information will bridge the knowledge gap and assist planning for adapting and mitigating climate change. Therefore, this chapter explains how agriculture, climate change, and the ecosystems interrelate among themselves in the main agroecological zones of Nigeria. Specifically, this chapter discussed the effects of the changing climate on farming calendar and local adaptation measures employed, the trend in primary drivers of social vulnerability to climate variability, the drivers of vulnerability to climate variation as perceived by farmers, health and environmental effects of climate instability on forest communities in Nigeria, and forest inhabitants needs in services and or facilities for an effective adjustment to climate change.

## Forest Livelihoods and the Challenge of Changing Climate

Forests are mainly situated in rural areas and/or frequently isolated areas (Aruwajoye and Ajibefun 2013). Apparently, such areas are in close harmony with nature with little development in terms of infrastructure (roads, potable water supply, markets, health facilities, and schools), government services, and jobs. Thus, it is not surprising that communities living in the fringe of these forests have limited livelihood opportunities and options (Wunder 2005). As a corollary to this, Bassey and Obong (2008) and Nayak et al. (2012) assert that communities around forest fringe are low-income earners who sought to build their economic capacity by engaging in the livelihood options provided by the forest.

In Nigeria, more than 90% of the rural populace depend on forests for their livelihood (Ayuk et al. 2011; Fadairo et al. 2017). These livelihood activities offered by forests include hunting of animals, forest-based farming, timber logging, gathering of building materials, collection of fuel wood for cooking or charcoal production, materials for local craft, fodder (grasses and leaves for livestock and grazing of livestock), medicinal plants, and non-timber forest products which include honey, leaves, and fruits. Suffice it to say that these forest products derived from the various livelihood activities are not solely for household consumption but also for commercial purposes.

Studies have documented the benefits accruing from forest resources in the livelihood activities of those inhabiting the fringes of forests as substantial (Levang et al. 2005; Sunderlin et al. 2005). These benefits according to Warner (2000) are increased income, improved food security, reduced vulnerability by providing safety nets, and increased well-being. Furthermore, some of these livelihood activities have social, religious, and cultural dimensions. For instance, hunting may serve as a cultural event for initiation into manhood while fishing maybe a social or cultural event. In addition to this, Shackleton (2004) opine that forests provide sites for spiritual healing and religious practices. Hence, it is not uncommon to find sacred places, herbalists, and native doctors in forest communities.

Agriculture as an important livelihood activity in most forest communities is affected by climate change in several ways, namely, changes in rainfall, standard temperatures, and climate extremes (heat weaves). Climate change influences planting and cropping conditions which in turn affects the supply of food. It necessitates changes in farming methods, increases soil pressure, reduces water supply to the root system, and increases farmers reliance on agrochemicals for farming. In addition, crops stressed as a result of climate changes become more susceptible to damage from diseases and pests. Animal husbandry industry is also indirectly affected following climate-induced changes in the availability of grains, pasture, and forages and its accustomed price increase. Animals health are usually affected negatively by extreme heat (Enete 2014). Furthermore, peasant and small-holder farmers who produce the bulk of food consumed in most developing countries are usually vulnerable to climate uncertainties due to their small size of farms, poor technology, and little working capital (Morton 2007). In addition, the seasonal calendar which provides information on planting, sowing, and harvestings periods of locally adopted crops in specific agroecological zones (Fadairo et al. 2019) is distorted by climate variability, predisposing farmers to risks arising from weather uncertainties.

## Agriculture, Climate Change, and Food Security

The occurrence of climate change is interlinked with the performance of agriculture and attainment of food security. Agricultural production in most parts of Africa including Nigeria has been seriously affected by environmental degradation caused by climate change, making a case for serious intervention (Osuafor and Nnorom 2014). The current global climatic condition has both natural and man-made causes. The reliance on rain-fed agriculture by most African countries both as a source of income and consumption has resulted in their high vulnerability to climate change. Some of the devastating effects includes: erosion, flooding, drought, pests and diseases, desertification, gas emissions, fluctuation in rainfall patterns, and a host of others. These factors in turn impact on agriculture and consequently threaten food security. As a result, food security is at risk with a daily world population increase. In order to forestall the danger ahead, the United Nations has clearly set the targets of attaining food surplus, food security, and improved nutrition, and advancing sustainable agriculture as number two among its 17 Sustainable Development Goals (SDGs) for the year 2030.

According to Food and Agriculture Organization (2002) as cited by Coates (2013), food security exists when all individuals, at every time, have socioeconomic and physical access to adequate, nutritious, and safe food that meets their dietary requirements and food preferences for a healthy and productive life. Therefore, food security goes beyond having adequate supply of food but also include issues relating to the food safety and hygiene. For instance, use of chemicals such as fertilizer in planting or produce preservation as a response to climate fluctuation predisposes the population to poor health. This assertion is in line with Kinsey (2005) who opined that a nation is not regarded as food-secure just because food is available in the right quantity needed by its populace, but also when the food consumption does not predispose the people to any health hazard. In order to reduce the impacts of climate change on agriculture, various coping strategies have been put in place. Osuafor and Nnorom (2014) highlighted the strategies as including controlling greenhouse gases emission; preventing deforestation; planting climate-smart, disease-tolerant, and high yield crops; and adjustment of planting calendars by farmers.

## Climate Change and Drivers of Social Vulnerability in Nigeria

In recent times, the variation in climate such as rise in temperature, increase in rainfall causing flood, delayed and inconsistent rainfall causing drought, strong wind, and landslides have threatened both the natural systems and the human society, specifically causing internal displacement of persons, destruction of lives, properties, and livelihood, food insecurity, disease outbreak, violence arising from struggle over resources, and increased suffering and penury. Yet, the impact caused by these climate extremes is not uniformly distributed among and within groups of people in the same country, state, and/or community (Petkova et al. 2015). Thus, it is unlikely for the impact of climate extremes to felt in the same way. Some groups or individual are likely to be more vulnerable than others. This underscores the need for researches to continually explore comparative analysis of climate change impacts across environmental, social, economic, and political factors in order to engender sustainable solutions.

In the views of Petkova et al. (2015), climate change effects will vary among age groups, sex, socioeconomic status, health condition, geographical location, and nature of livelihoods of the people. A measure of the extent of exposure of groups or individuals to stress as a result of the impacts of climate change extremes is known as social vulnerability. In this chapter, livelihood of the forest-edge communities in Nigeria is singled out among others for discussing the effects of vulnerability to climate change. Hence, stress in this perspective refers to the interference of the livelihood activities of groups or individuals in the face of climate extremes.

Evidence abounds that the people who are likely to be more susceptible to the adverse effects of climate extremes are people in the rural areas. Rural Nigeria is mainly agrarian with many of them living below the poverty line. Research reveals that rural communities are inexplicably vulnerable to climate extremes because their livelihoods are dependent on climate-sensitive activities (agriculture, forestry, fishing, recreation) in their rural environment (Fisher et al. 2013). Therefore, the effect of climate extremes poses a huge threat especially to the agrarian rural people many of whom already live below the poverty line.

## Theoretical Framework

The Sustainable Livelihood Framework, as presented below, was considered relevant for underpinning the assumptions and approach utilized in this chapter.

### Sustainable Livelihood Framework

Sustainable livelihood (SL) framework presents a tool for development workers to understand, analyze, and explain the real factors that affecting poor people's livelihood (Petersen and Pedersen 2010). According to Carloni and Crowley (2005), a livelihood comprises the assets (including both social and material endowments), capabilities, and activities necessary to earn a living. Ability to manage and recover from shocks and stresses, retain or enhance its assets and capabilities, while not depleting the natural resource base is what makes a livelihood sustainable. SL framework addresses the creation of guaranteed livelihoods for the poor by development workers. The basic principle of SL is that development work ought to focus on the people, with considerations of, what matters for the poor, cultural diversity, and its effects on livelihood processes. Secondly, poor people themselves are major actors in bringing about the change they desire. This is because they have a better knowledge of issues affecting them much more than any external person (Petersen and Pedersen 2010). The foregoing reechos the central argument in this chapter that adequate understanding of the perspectives of the local people whose livelihoods are

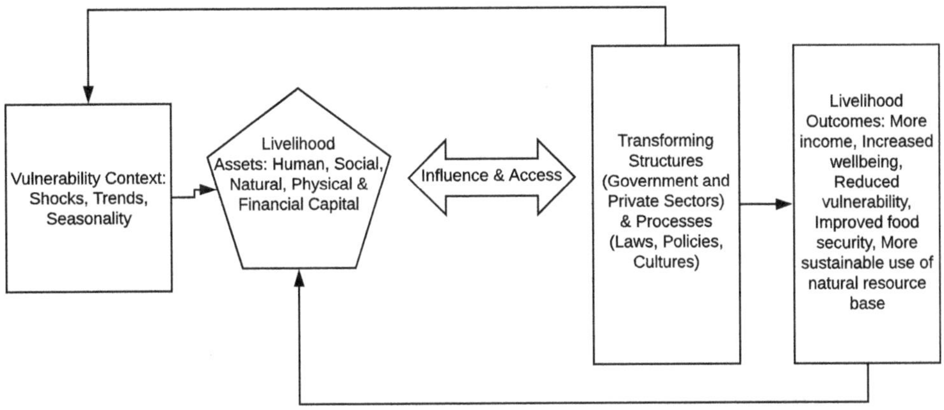


**Fig. 1** Sustainable Livelihood Framework. (Source: DFID (2000), cited in Petersen and Pedersen (2010))

intertwined with and affected by climate variability is important for a more sustainable adaptation and mitigation measures for climate change. Hence, primary information used for discussion in this chapter were derived from engagement with those who are most affected in order to acquire their perspectives of their problems and what things to change to improve their condition. The key components of the SL framework are indicated in Fig. 1, and they include vulnerability context, livelihood assets, structure and process for transformation, livelihood strategies, and livelihood outcome.

**Vulnerability context**: This refers to people's external environment. It includes occurrences for which people have restricted or no control. Examples of such occurrences are critical trends of economic inflation, natural disasters, shocks, and seasonality, among others. The issue of vulnerability thus emerges when individuals are exposed to harmful threats they are not well equipped to confront (Petersen and Pedersen 2010).

**Livelihood assets:** Since the framework by nature focuses on the people, it thus seeks to have a better consideration of the people's power (capitals or assets). Since the approach relies on the belief that achieving livelihood outcomes requires a combination of assets, understanding how the conversion of the people's power to favorable livelihood outcomes becomes paramount. For this reason, the framework identifies five forms of capitals which support livelihoods. These are social, human, natural, financial, and physical capitals (Petersen and Pedersen 2010).

**Structure and process for transformation**: This refers to the establishments and regulations found from the household to the international levels that defines the livelihoods of the poor. These establishments and regulations stimulate how people access the various types of assets. Ownership rights and laws to secure individual rights are examples of processes, whereas structures are things like existence of ministries, banks with credit facilities for support groups and farmers (Petersen and Pedersen 2010).

**Livelihood strategies:** This refer to the way in which people organize towards achieving their anticipated livelihood and access to diverse types of resources determine the approaches to be employed. Furthermore, societal opportunities or constraints can be dictated by its structures and processes.

Finally, **livelihood outcome** refers to the resultant effect of livelihood strategies assumed by the people. These effects could be better income, enhanced wellness, decreased susceptibility, and food security, among others (Petersen and Pedersen 2010).

## Chapter Approach

This chapter utilized synthesis of literature and collection of primary data to reach its conclusions. The field activities for primary information garnering were carried out in Nigeria. Nigeria's ecological environment consists of seven agroecological zones such as saltwater swamp, freshwater swamp, Sudan savannah, guinea savannah, Sahel savannah, tropical rainforest, and the montane zones. However, discussions in this chapter are mainly focused on its three broadly classified agroecological zones namely savannah, mangrove/swamp, and rainforest. Each of these agroecological zones has their own peculiarities and supports a wide range of plant and animal species. Nevertheless, the tropical rainforest has been adjudged the richest. All adult residents in forest communities who are engaged in farming and or other forest-based livelihood activities were engaged in discussions. Adults who were 45 years and above at the time of the field work were specifically targeted due to the 30 years reference period used in this chapter. A two-stage sampling procedure was used. First, from each of the three major agroecological zones, one state was purposively sampled based on distribution and extent of forest cover. Second, two forest communities in each sampled state were selected purposively based on intensification of climate variability in the past 15 years. Thus, six forest communities namely Iyamitet, Ikom Agoi (Cross River State), Wawa-Gbere, Emi-Hakimi Mokwa (Niger State), Osoku, and Fowowa (Ogun State) were sampled for primary information garnering (Fig. 2). In each of the sampled locations, a short survey was carried out to generate a pool of potential participants for focus group discussions as follows:

> Would you be willing to participate in a focus group discussion regarding climate change, rural livelihoods and other related issues? The discussion would take about 2-3 hours on ........... (date) and you would be incentivised for participation. The discussion will be audiotaped for the purposes of review by the researchers.
>
> Yes [ ] No [ ]. Name............... Phone no. ............
>
> Age.......................

Given the size of potential participants generated, four focus group discussions comprising 10–12 members per group were held in each forest community. Thus, qualitative data were collected from 24 focus group discussions held with 252 male and female farmers in selected sites. Deforestation and Forest Degradation Analysis adapted

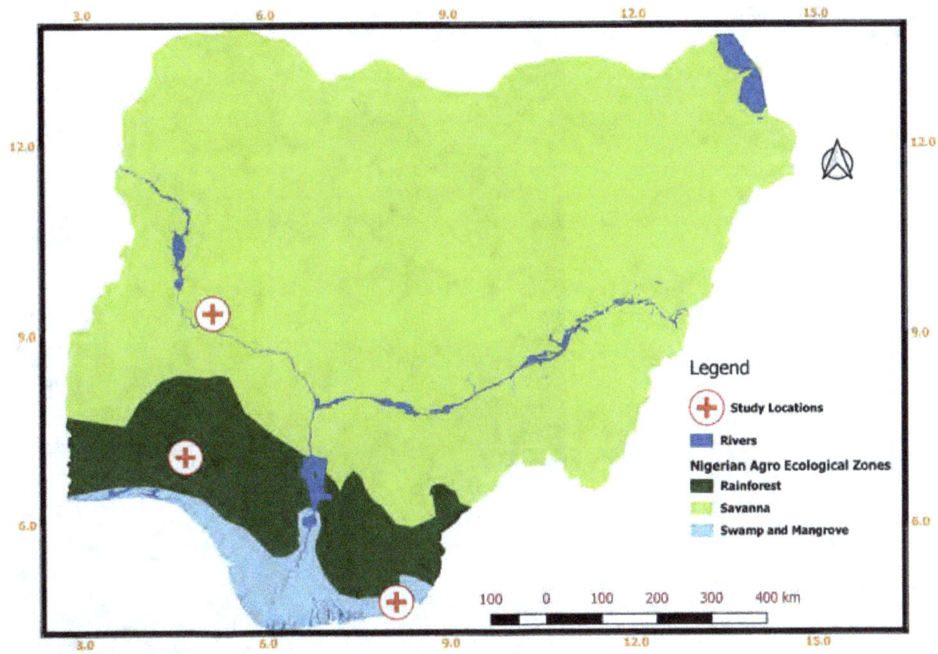


**Fig. 2** Map showing the focus group discussion locations within the agroecological zones of Nigeria

from Tiani et al. (2015) was used in this chapter. Using a discussion guide and 30 years as reference, primary information was sought on effects of climate change on farming calendar, trends in primary drivers of social vulnerability to climate change, causes and consequences of environmental degradation in forest communities, and communities needs in services and/or facilities for an effective adaptation.

A visual representation showing seasonal activities among community members on a flip chart was used to facilitate conversation on how climate change affects farming calendar. Discussants indicated their cropping pattern in a calendar year, the associated activities, and why the activities are conducted in order to provide information on farmers' local climate change adaptation measures. Also, participants during the discussions used pebbles (stones) to proportionately represent areas covered by each ecosystem components (population, agriculture, forest cover, hunting, and charcoal production) as perceived over the past 30-year period (Figs. 3, 4, 5, and 6). The outcome of this exercise was represented in a single table to indicate the trend as perceived by the discussants. Influence of climate change on local livelihood activities were captured by asking participants to award scores to represent magnitude and impact of each climate parameter on available livelihood activities. Problem tree analysis and paired needs ranking participatory tools were used to investigate health and environmental impacts of unstable climate on forest communities and communities needs in services and or facilities for an effective adaptation to climate change, respectively.

During the discussions, audiotape recording, flip chart, and handwritten notes were taken by researchers and were later transcribed. Primary information collected

**Fig. 3** Male participants during focus group discussion at Mokwa, Niger State on March 9, 2018

were coded and analyzed based on thematic patterns in the participants responses to issues raised in the focus group discussions. In doing this, particular attention was paid to important quotes from some of the respondents. Some of these quotes are used as evidences to support discussions in this chapter.

Limitations however exist in the approach used in this chapter due to language barrier experienced in some of the field locations as we observed that some respondents could not communicate well in the English language. Therefore, the researchers relied on translators to interpret the questions to the respondents and participants' responses back to the researchers. Some content and meaning may have been lost in this process. Also, due to the problems of insurgency, farmer-herdsmen conflict, and other security issues, most states within the savanna agroecological zone were deliberately excluded from the focus group discussions.

## Effects of Climate Variability on Cropping Calendar in Mangrove, Rainforest, and Savannah Agroecological Zones of Nigeria

Farm households' ability to grow enough food to feed themselves and their animals is determined to a large extent by the weather since agricultural production depends on climate variables, such as temperature, precipitation, and light. Therefore,

**Fig. 4** Male participants during focus group discussions at Iyamitet, Cross River on 14 February, 2018

**Fig. 5** Female participants at FGD at Mokwa on March 9, 2018

**Fig. 6** Female participants at FGD on April 11, 2018 at Osoku

shifts in temperature and precipitation are important parameters for farming communities in the timely operations of farm operations. Table 1 shows the activities done in relation to the major crops produced by each of the ecological zones. Yam, cocoa, and maize were focal crops in the mangrove, rainforest, and savannah zones, respectively. Within the mangrove ecological zone, the major activity between January and May was land preparation, with dry and wet seasons being observed in March 30 years ago. Currently, there is a little shift, as dry and wet seasons are now being observed in April. It was clear that there is a relatively longer dry season now compared to what it was 30 years ago as wet season was fully experienced from April to September, lasting a period of 6 months. In the recent times, however, the wet season now occur in May to August, lasting only 4 months. This trend is similar to what is observed in the rainforest and savannah zones with extended dry season compared with the referenced 30 years ago. This shows a drop in the duration of rainfall across the three agroecological zones. This change in climatic trend is in line with the position of Appiah et al. (2018) that there is a decrease in the intensity of rain in most forest reserves communities of most sub-Saharan African countries where there is still heavy reliance on agriculture as primary means of livelihood. This change has implications for agriculture in Nigeria which is known to be mainly rain-fed.

Further, two of the participants at focus group discussions gave explicit distinctions of the season and farming activities between 30 years ago and now as thus:

**Table 1** Seasonal and cropping calendar for major crops in mangrove, rainforest, and savanna ecological zones of Nigeria

| | Months of the year (January-December, respectively) | | | | | | | | | | | |
|---|---|---|---|---|---|---|---|---|---|---|---|---|
| | **J** | **F** | **M** | **A** | **M** | **J** | **J** | **A** | **S** | **O** | **N** | **D** |
| **Mangrove** | | | | | | | | | | | | |
| Season 30 years ago | D | D | D/W | W | W | W | W | W | W | D/W | D | D |
| Season currently | D | D | D | D/W | W | W | W | W | D/W | D | D | D |
| Activities currently | | | | | | | | | | | | |
| • Land preparation | ■ | ■ | ■ | ■ | ■ | | | | | | | |
| • Planting | | ■ | ■ | ■ | ■ | ■ | | | | | | |
| • Weeding | | | | | ■ | ■ | ■ | ■ | ■ | ■ | | |
| • Staking | | | | | ■ | ■ | ■ | | | | | |
| • Harvesting/storage | ■ | ■ | | | | | | | ■ | ■ | ■ | ■ |
| **Rainforest** | | | | | | | | | | | | |
| Season 30 years ago | D | D | D/W | W | W | W | W | W | W | W | D/W | D |
| Season currently | D | D | D | D/W | W | W | W | W | W | DW | DW | D |
| Activities currently | | | | | | | | | | | | |
| • Land preparation | ■ | ■ | ■ | | | | | | | | | |
| • Cocoa nursery | | | ■ | | | | | | | | | |
| • Planting | | | | ■ | ■ | | | ■ | | | | |
| • Weeding | | | | | ■ | ■ | ■ | ■ | ■ | | | |
| • Spraying | | | | | | ■ | ■ | ■ | ■ | ■ | | |
| • Harvesting/storage | | | | | | | | | ■ | ■ | ■ | ■ |
| **Savanna** | | | | | | | | | | | | |
| Season 30 years ago | D | D | D/W | W | W | W | W | W | W | W | W/D | D |
| Season currently | D | D | D | D | W | W | W | W | W | W/D | D | D |
| Activities currently | | | | | | | | | | | | |
| • Land preparation | ■ | ■ | ■ | | | | | | | | | |
| • Planting | | | | ■ | ■ | ■ | ■ | ■ | | | | |
| • Weeding | | | | | | | ■ | ■ | ■ | | | |
| • Harvesting/storage | | | | | | | | | ■ | ■ | ■ | ■ |

*D =Dry, W = Wet, D/W =Dry & Wet*

> Thirty years ago, we experienced little rainfall in the month of March and then the rains become fully established in April and then fade away in September. Presently, we experience little rainfall in the month of February which become fully established in April and then fades out in October.
>
> Land clearing of farmland in our area has shifted from January/February to March. This happens because onset of rain has shifted to April, currently any rain we see in February/March is tagged accidental rain.

Another discussant also said:

> ‘Compared to thirty years ago, there is reduced intensity and duration of rainfall’ (A 50-year-old man, Muslim, farmer, savannah zone).

Access to food through both production and exchange will continue to depend not only on the productivity and profitability of agriculture, but also on how well the political climate enables people to respond creatively to their environment and prospects. As a means of adapting to the extended dry spell, farmers have adopted the cultivation more hardy crops such as cassava to reduce economic losses associated with climate change. Diversification into trading and processing of agricultural produce also take place during the off seasons to bridge income gap that is now experienced. This corroborates the assertion of Appiah et al. (2018) which argues that farmers in forest communities are often engaged in other economic activities to supplement their agriculture-based incomes. One of the discussants said thus:

> Presently, when there is no planting activity, we engage in other businesses like buying and selling so as not to be idle and to earn money.

## Perceived Trend in Agriculture, Forest Area, and Population as Primary Drivers of Social Vulnerability

The relationship between climate change and ecosystem is intertwined. Climate change can affect the distribution and behavior of some ecosystem components. Conversely, the intensification of some ecosystem components such as forest cover also have implications for carbon sequestration capacity, and hence extent of climate change impact on the environment. In this chapter, element of the ecosystem such as agriculture, forest cover, population, and alternative income-generating activities in the sampled sites such as fishing, hunting, and charcoal production were focused for understanding the interrelationship among these elements and climate change. Across the three agroecological zones of rainforest, mangrove, and savanna, more people are presently involved in agricultural activities compared with 30 years ago. This, according to discussants, is due to lack of employment for graduates and irregularities in payment of salaries/wages to government workers. Therefore, the need for alternative sources of income or employment (for the unemployed) has caused the recent surge in the farming population in the areas. A 48-year-old woman farmer in one of the rainforest communities explained thus:

> Very few people were involved in agriculture thirty years ago in our community, then, youth only engaged in land clearing, planting and weeding of their parents' farms. However, as our population increases many people got involved in agriculture and young people now cultivate their own farmlands.

Population as a component of the ecosystem also witnessed increasing trend in the last 30 years in all the agroecological zones. Population is a key factor that differentiate Africa from other regions of the world. African population is projected to grow rapidly throughout the twenty-first century and this growth will have direct effect on the demands for agricultural commodities. Discussants explained that increase in population is caused mainly by two factors. First, migration of people into various communities within the area for livelihood, especially in agriculture. Rural areas are known to be rich in an important factor of production which is land. Second factor was increase in child-bearing and survival of children due to improved health awareness. However, while there was increase in population and intensity of agriculture across the agroecological zones, decrease in forest spread and abundance of non-timber forest products such as snails and mushrooms was observed across the zones in the last 30 years. This is caused by deforestation and bush burning, occasioned by increase in population. Increasing population exerts more pressure on resources leading to upsurge in the rate of felling of trees for cooking and charcoal production. Increasing involvement of people in agriculture and declining soil nutrient also led to agricultural "extensification" and the consequent expansion of land for agricultural activities. The implication is that more forest cover is removed in order to increase size of land cultivated for farming. A 52-year-old man in one of the savannah communities remarked during the focus group discussions as follows:

> The forest cover has drastically reduced against how it was. In those days we usually have 10-2 stumps of tree together, but due to deforestation by saw millers, increase in land acquisition for farming activities and construction of houses, almost all the trees are gone.

Further, hunting and fishing activities have assumed a downward trend in the savanna, mangrove, and rainforest zones in the last 30 years. This can be attributed to a decline in their distribution in their natural habitat and hence, reduced motivation on the part of the hunters and fishermen to continue in the business due to low rewards. In addition, reduced trend in fishing activities in the locations is plausibly due to water pollution, overfishing, and reduction in water volume caused by prolonged drought and a decline in intensity and duration of rainfall. Idowu et al. (2011) confirmed a decline in Catch Per Unit Effort (CPUE) in coastal areas of Nigeria due to pressure of climate change. However, charcoal production is among the gradually becoming prominent livelihood activities in savannah and mangrove zones. This is perhaps as a replacement for fishing and hunting livelihood activities that are already becoming faded in the areas. Charcoal production serves as an alternative income source for farmers in rural communities while ignoring its long-term implications for a sustainable environment. This is consistent with Mwampamba (2007) who had projected that by 2028, public forest resources would be depleted in some parts of Africa and there would be a total collapse of charcoal chain if no measures are put in place to stop the trend. Mwampamba's position suggests that there is an arbitrary involvement in charcoal production and that the rate of growth in the sector is largely unsustainable. Massive charcoal production in the areas have implications for availability of None Timber Forest Products (NTFPs) in the area. During the focus group discussion, one of the discussants explained (Table 2):

> Duration of time used for collecting products in the forest has reduced because of scarcity of these products in the forest. Harvest from NTFPs was much thirty years ago (A 45-year-old woman, Christian, farmer, rainforest zone)

**Table 2** Trend in ecosystem components as primary drivers of social vulnerability to climate change

| | | Index of intensification in last 30 years (%) | | | |
|---|---|---|---|---|---|
| Agroecological zones | Ecosystem components | 1987 | 1997 | 2007 | 2017 |
| ***Rainforest*** | Agriculture | 18.6 | 21.0 | 24.6 | 36.0 |
| | Population | 25.0 | 20.6 | 21.6 | 33.6 |
| | Forest cover | 44.6 | 24.6 | 19.6 | 11.6 |
| | Fishing | 58.0 | 22.0 | 14.0 | 6.0 |
| | Charcoal production | 100 | 0 | 0 | 0 |
| ***Mangrove*** | Agriculture | 18.5 | 21.0 | 24.5 | 36.0 |
| | Population | 25.0 | 20.0 | 21.5 | 33.5 |
| | Forest cover | 44.5 | 24.5 | 19.5 | 11.5 |
| | Hunting | 56.0 | 24.0 | 14.0 | 6.0 |
| | Charcoal production | 92.0 | 0 | 0 | 8.0 |
| ***Savanna*** | Agriculture | 21.0 | 15.5 | 20.5 | 43.0 |
| | Population | 15.0 | 20.0 | 23.0 | 42.0 |
| | Forest cover | 46.0 | 24.0 | 18.0 | 12.0 |
| | Fishing | 38.0 | 20.0 | 20.0 | 22.0 |
| | Charcoal | 4.0 | 9.0 | 19.0 | 68.0 |

## Perceived Effects of Drivers of Climate Change on Household Livelihood

Weather patterns are becoming unpredictable due to increasing variability in climate parameters. Rising temperature and increased frequency of extremely dry and wet years are expected to slow progress in crop productivity, livestock system, and improved food security. This section explains the effects of drought, flood, pest and diseases, increase in temperature, and strong wind as observable drivers of vulnerability to climate change in the Nigeria agroecological zones. The effect of each was captured in both direction (negative or positive) and magnitude (which measures the extent of such effects) on subsistent crop, cash crop, charcoal production, and animal production. In the agroecological zones, the negative effects of drought were most felt on subsistence crops. Cash crops such as cocoa were not spared as the effect was high, especially on the field. An opinion leader among the farmers in Iyamitet community (mangrove zone) explained thus:

> In recent times, there is incessant invasion of pests and diseases on our crops which we cannot effectively control. Most times, it takes a lot of time before we understand the nature and causes of such infestation and able to adjust to it. The result is a serious reduction in the quality of our crops and huge losses for us. Ugly experience from the last year harvest has even discouraged some few cocoa farmers from business.

Discussions further revealed that charcoal production was the most resistant or the least affected by drought in the ecological regions. The reverse was the case with respect to flooding, as charcoal production was reportedly highly affected in terms of quality of charcoal, quantity of wood (log), and duration of activities of charcoal production. The level of vulnerability was also high for subsistent crops, cash crops, and availability of NTFPs. Discussants during the focus group sessions justified their position that flood has a high negative effect on charcoal production explaining that when flood occurs, it washes away the charcoal heaps and woods, thereby terminating the charcoal production process. The implication is that while flooding is not an everyday occurrence, the few times it occurs, it has considerable negative effect on the quantity, quality, and hence profitability of charcoal production among producers.

Furthermore, livestock production was the least threatened livelihood by increased temperature, with little or no observable implications for livestock/milk production. However, almost all the livelihood activities were threatened by increased temperature with negative implication for future and present income generation among the people. The effects of pests and diseases, which were also directly associated with climate change as reported earlier, is also evident on subsistent crop, NTFPs, and cocoa production. This rating agrees with the position of majority during discussions as they were unanimous that pests and diseases had grown in both frequency of occurrence and severity of effects in the past few years with observable negative effects on crops and other means of income generation. On the other hand, the effects were less felt on charcoal and livestock production.

The last driver considered was strong wind. This chapter reveals that strong wind had the least effect on respondent's various livelihood activities, with the most visible effects on NTFPs and the least on livestock production. The effect was rated moderate on charcoal production. This rating was backed by some explanations during discussion as follows:

> Strong wind often has negative effect on quantity of charcoal produced. This is often due to strong wind and availability of opening on the charcoal heap. This causes air to enter the charcoal heaps. This occurrence often reduces production efficiency, leading to high ash to charcoal ratio, thereby reducing quantity and profitability of our charcoal production enterprise.

## Causes and Consequences of Environmental Degradation in the Nigeria Agroecological Zones

The problem tree analysis in Fig. 7 illustrates the composite results of the effects of environmental degradation in the three ecological zones. It gives the causes and effects of the common problem identified as environmental degradation. The causes of environmental degradation identified in this section are consistent with earlier discussions and include drought, deforestation, strong wind, delayed rainfall, bush burning, use of agrochemicals, and air pollution. This corroborates the report of Somorin (2010) that impact of climate change could be felt by increased temperature, deforestation, and drought. Discussants also reported that the effects of

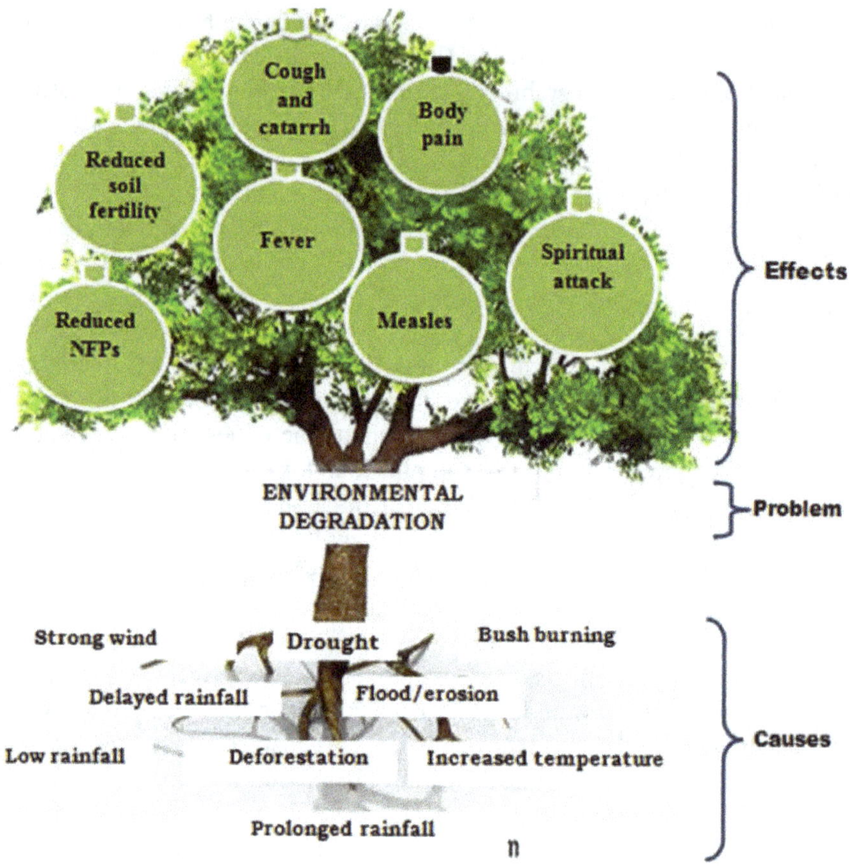


**Fig. 7** Problem tree analysis of causes and consequences of environmental degradation in Nigeria

environmental degradation produced both adverse health and environmental consequences such as mental disorder, measles, meningitis, water scarcity, erosion, cough/catarrh, body pain, reduced NTFPs activities, erosion, poverty, low income and poor harvest, soil infertility, increased incidence of pest and diseases and fever.

Specifically, in savannah zone, respondents experienced strong wind which is carries particles of dust causing cough and catarrh. Idowu et al. (2011) established that respiratory diseases and infections like cough and catarrh are prominent in harsh climate due to presence of pollutants and dust. Also, increased temperature in the area causes meningitis and measles which is very common in children of less than 5 years. A 37-year-old woman from savannah zone and 30-year-old lady from mangrove zone are quoted, respectively, as follows while commenting on the consequences of some of climate change parameters on their health and livelihoods:

> During hot weather, as we now frequently experience here, there is usually the prevalence of meningitis, measles and chicken pox especially among our children. Also, wind is accompanied with dust, and this brings about occurrence of cough and catarrh is common. Increased temperature also affects our animals and crops as pests and diseases grow more

> when the weather is hot which slow down crop production (A 37-year-old woman, Muslim, farmer, savannah zone).
>
> The impact of wind is low and it could be beneficial when drying crops after harvesting them. But when it is too strong, it could destroy the crops' (A 30-year-old woman, Christian, processor, mangrove zone)

An interesting link between deforestation and pervasiveness of mental disorder among the people was alluded to in Wawa community (savannah zone) where the participants argued that deforestation releases certain spirits which are believed reside in trees to inflict young ladies with insanity. One of the male discussants said the following:

> Due to deforestation, spirits that abode in the trees are made homeless and therefore come to town and enter into our young ladies making them go insane. We are able to establish this because during deliverance session for some of the victims, the spirits confessed that their natural habitats have been disrupted and that's the reason for the attacks.

## Farming Communities' Needs for an Enduring Adaptation to Climate Change in Nigeria

Farming communities' adaptation is key in translating climatic challenges and agricultural responses into changes in production, prices, food supply, and welfare. The potential for positive change for farming communities will increase if farmers are helped to adapt to climate variability. Table 3 therefore show the priority ranking of what communities in the agroecological areas adjudged as their needs to effectively adjust to climate changes impact. Across the ecological zones, respondents highlighted some needs that were important due to various problems and setbacks they encountered arising from climate change. Availability of credit facilities ranked first, suggesting lack of sufficient capital for various livelihood activities and effective climate change adaptation. Building of health center

**Table 3** Needs matrix as identified by communities in forest-edge communities in Nigeria

| Needs | Ranks |
|---|---|
| Credit facilities | 1 |
| Heath center | 2 |
| Farm inputs – improved seed variety, fertilizer, herbicides | 3 |
| Good road network | 4 |
| Pipe borne water | 5 |
| Power supply | 6 |
| Irrigation | 7 |
| Communication network | 8 |

ranked second as most of the communities' lacked access to well-equipped health centers that can provide health care services. Farmers also require farm inputs in the form of herbicides, improved seed varieties, and fertilizer as the third ranked need. Perhaps, the need to easily connect outside communities for timely marketing of agricultural produce necessitated farmer's choice of good access-linking road. Discussions revealed that farmers find it difficult to sell their produce to right buyers due to poor state of farm-to-market road. Instead, commodities were often sold to middle men at farm gate prices and hence, low profitability for the farmers. Other infrastructure such as pipe borne water, power supply, communication network was also mentioned as some of their pressing needs across communities in the agroecological zones. The huge infrastructural deficit in the sampled locations is indicative of the state of physical development in most of the forest-edge communities in Nigeria. This corroborates the earlier assertion by one of the discussants:

> 'We strongly desire credit facilities and good road network to enable us perform better in our farming activities and also for outsiders to come into our community and trade with us' (A 52-years-old man, Christian, farmer, rainforest zone)

## Conclusion and Recommendation

This chapter concludes that the scale and direction of climate change impact on agriculture as the primary rural livelihood and other ecosystem components in Nigeria's agroecological zones is largely comparable and have both positive and negative consequences on rural sustenance. While climate change impact combined with some other economic factors such as unemployment have encouraged urban-rural migration, agricultural intensification, and livelihood diversification on the one hand, it has increased vulnerability tendencies of rural households in forest-edge communities in all the agroecological zones of Nigeria on the other hand. Among several others, increased forest encroachment, lack of a well-coordinated policy framework which allows for alternative livelihood without accompanying forest regulatory framework were major vulnerability exacerbating factors for the rural poor. Although the rural populace need help for better adjustment to climate change, they also do have demonstrated ability to respond to changes occasioned by climate variability and are exploring these abilities to the best of their knowledge. Support in the form of affordable insurance policy, credit, agri-inputs, favorable forest regulatory framework, and youth empowerment would enhance sustainable adjustment to climate change among the rural people.

**Acknowledgments** This work was supported by the African Development Bank and Japan Trust Fund under the Education for Sustainable Development in Africa (ESDA) project on knowledge creation, sharing, and exchange between Asia and Africa in support of Sustainable Development of Africa.

## References

Appiah DO, Akondoh ACK, Tabiri RK, Donkor AA (2018) Smallholders farmers' insight on climate change in rural Ghana. Cogent Food and Agriculture 4:1–16. http://www.congentoa.com/article/10.1080/23311932.208.1436211.pdf/27/April/2018. Assessed 17 June 2018

Aruwajoye AS, Ajibefun IA (2013) Assessment of livelihood strategies among households in forest reserve communities in Ondo state, Nigeria. Int J Sci Eng Res 4(7):1650–1655

Ayanlade A, Radeny M, Morton JF, Muchaba T (2018) Rainfall variability and drought characteristics in two agro-climatic zones: an assessment of climate change challenges in Africa. Sci Total Environ 630:728–737

Ayuk ET, Fonta WM, Ichoku HE (2011) Forest income and household welfare in rural Nigeria. https://unu.edu/publications/articles/forest-income-and-household-welfare-in-rural-nigeria.html#info. Accessed 20 Nov 2019

Azeez IO, Ikponmwonba OS, Popoola L, Amusa TO (2010) Land use activities among forest environments' dwellers in Edo State, Nigeria: implications for livelihood and sustainable forest management. Int J Soc For 3(2):164–187

Bassey BJ, Obong LB (2008) The role of non-timber forest products (NTFR) in community livelihoods: a case study of Okiro, Obudu Local Government Area Cross River state, Nigeria. J Environ Res Policies 3:98–105

Carloni AS, Crowley E (2005) Rapid guide for missions: analysing local institutions and livelihoods. Food and Agriculture Organization of the United Nations (FAO), Rome

Coates J (2013) Build it back better: deconstructing food security for improved measurement and action. Global Food Secur 2(3):188–194

Cooper PJM, Dimes J, Rao KPC, Shapiro B, Shiferawa B, Twomlow S (2008) Coping better with current climatic variability in the rain-fed farming systems of sub-Saharan Africa: an essential first step in adapting to future climate change. Agric Ecosyst Environ 126:24–35

Vidaurre de la Riva M, Lindner A, Pretzsch J (2013) Assessing adaptation – climate change and indigenous livelihood in the Andes of Bolivia. J Agric Rural Dev Trop Subtrop 114(2):109–122

Enete IC (2014) Impacts of climate change on agricultural production in Enugu State Nigeria. J Earth Sci Clim Change. https://doi.org/10.4172/2157-7617.1000234

Fadairo OS, Calland R, Mulugetta Y, Olawoye JE (2017) A corruption risk assessment of reducing emissions from deforestation and forest degradation project in Nigeria. Int J Clim Change Impact Responses 10(1):1–22

Fadairo O, Williams PA, Nalwanga FS (2019) Perceived livelihood impacts and adaptation of vegetable farmers to climate variability and change in selected sites from Ghana, Uganda and Nigeria. Environ Dev Sustain. https://doi.org/10.1007/s10668-019-00514-1

Fisher PA, Paveglio T, Carroll M, Murphy D, Brenkert-Smith H (2013) Assessing social vulnerability to climate change in human communities near public forests and grasslands: a framework for resource managers and planners. J For 111(5):1–9. https://doi.org/10.5849/jof.12-091

Idowu AA, Ayoola SO, Opele AI, Ikenweiwe P (2011) Impact of climate change in Nigeria. Iranica J Energy Environ 2(2):145–152

Kinsey J (2005) Will food safety jeopardize food security? Agr Econ 32(s1):149–158

Levang, P., Dounias, E. & Sitorus, S. (2005). Out of the forest, out of poverty? Forests, trees and livelihoods, 15(2): 230

Lindner A, Pretzsch J (2013) An international network on climate change impacts on small farmers in the tropical Andes – global conventions from a local perspective. Sustain Agric Res 2 (2):92–98. https://doi.org/10.5539/sar.v2n2p92

Malhi Y, Wright J (2004) Spatial patterns and recent trends in the climate of tropical rainforest regions. Philos Trans Biol Sci 359:311–329

Morton JF (2007) The impact of climate change on smallholder and subsistence agriculture. PNAS 104(50):19680–19685

Mwampamba TH (2007) Has the woodfuel crisis returned? Urban charcoal consumption in Tanzania and its implications to present and future forest availability. Energy Policy 35 (8):4221–4234

Nayak BP, Kohli P, Sharma JV (2012) Policy brief: livelihoods of communities and forest degradation in India: issues for REDD$^+$ policy. Https://www.semanticscholar.org/paper/Livelihood-of-local-communities-and-forest-in-India-Nayak-Kohli/4a45c80a48147a4c607b96905212d6bf5f5622f6. Accessed 17 Nov 2019

Osuafor AM, Nnorom NR (2014) Impact of climate change on food security in Nigeria. Int J Sci Technol 3(1):1–16

Petkova EP, Ebi KL, Culp D, Irwin R (2015) Climate change and health on the U.S. Gulf Coast: public health adaptation is needed to address future risks. Int J Environ Res Public Health 12:9342–9356

Petersen EK, Pedersen ML (2010) The sustainable livelihoods approach: from a psychological perspective. Institute of Biology, University of Aarhus

Power AG (2010) Ecosystem services and agriculture: trade-offs and synergies. Philos Trans R Soc 365:2959–2971. https://doi.org/10.1098/rstb.2010.0143

Pretty J, Toulmin C, Williams S (2011) Sustainable intensification in African agriculture. Int J Agric Sustain 9(1):5–24

Saka MG, Jatau DF, Olaniyi WA (2013) Status of indigenous tree species in Girei forest reserve of Adamawa State. J For Res Wildlife Environ 5:28–40

Shackleton CM (2004) Assessment of livelihoods importance of forestry, forests and forest products in South Africa. Rhodes University, Grahamstown

Somorin OA (2010) Climate impacts, forest-dependent rural livelihoods and adaptation strategies in Africa: a review. Afr J Environ Sci Technol 4(13):903–912

Stringer LC, Dougill AJ, Fraser E, Hubacek K, Prell C, Reed MS (2006) Unpacking "participation" in the adaptive management of social–ecological systems: a critical review. Ecol Soc 11(2):39. http://www.ecologyandsociety.org/vol11/iss2/art39/. Assessed 12 July 2018

Sunderlin W, Angelsen A, Belcher B, Burgers P, Nasi R, Santose L, Wunder S (2005) Livelihoods, forests and conservation in developing countries: an overview. World Dev 33(9):1383

Tiani AM, Besa MC, Devisscher T, Pavageau C, Butterfield R, Bharwani S, Bele M (2015) Assessing current social vulnerability to climate change: a participatory methodology (Working paper no. 169). Center for International Forestry Research, Bogor. ISBN 978-602-1504-70-3. https://doi.org/10.17528/cifor/005463

Tompkins EL, Adger WN (2004) Does adaptive management of natural resources enhance resilience to climate change? Ecol Soc 9(2):10. http://www.ecologyandsociety.org/vol9/iss2/art10. Assessed 17 Aug 2018

Veenendaal EM, Swaine MD (1998) Limits to tree species distribution in lowland tropical rainforests. In: Newbery DM, Prins HHT, Brown M (eds) Dynamics of tropical communities. Blackwell Publishing, Oxford

Ward RD, Friess DA, Day RH, Mackenzie RA (2016) Impacts of climate change on mangrove ecosystems: a region by region overview. Ecosyst Health Sustain 2(4)

Warner K (2000) Forestry and sustainable livelihoods; what part can forest and forestry play in reducing poverty? Unsilver: FAO, 51(202), 3–12

Wunder S (2005) Forestry and sustainable livelihoods. World Commission on Environment and Development (1999). Our Common Future. Oxford University Press, pp 1817

# 10

# Social Vulnerability of Rural Dwellers to Climate Variability: Akwa Ibom State, Nigeria

Jemimah Timothy Ekanem and Idongesit Michael Umoh

## Contents

Introduction .
Rural Population Characterization in Akwa Ibom State
Social Vulnerability of Rural Dwellers to Climate Variation in Akwa Ibom State
Factors Affecting Social Vulnerability to Climate Variation .
Empirically Defining the Underlying Dimensions of Factors Affecting Social Vulnerability .
- Local Support Factor .
- Educational Challenge Facto
- Income /Employment/Replacement Ability Factor .
- Family Structure Factor .
- Demographic Explosion Factor
- Age of Resident Factor
  - Gender Inequality Factor
  - Tribal Issues, Public Assault and Insecurity Factor
  - Infrastructure Inaccessibility Factor .
  - Loss of Properties/ Accommodation Factor
  - Mobility Factor
  - Disability Factor
  - Weak Local Institution Facto
  - Cultural Constraints Factor 6
  - Climate Information/ Communication Factor

---

J. T. Ekanem (✉)
Department of Agricultural Economics and Extension, Faculty of Agriculture, Akwa Ibom State University, Uyo, Nigeria
e-mail: jemimahekanem@aksu.edu.ng

I. M. Umoh
Agricultural Science Education Unit, Department of Science, Redemption Academy, Uyo, Nigeria

Loosed Social Network and Safety Net Factor
Environmental / Social Discrimination Facto
Measures to Reduce Rural Residents' Social Vulnerability to Climate Change
Conclusio
Recommendation
References

**Abstract**

For their livelihood activities, rural farming communities depend more on extractive capital. Their capacity to cultivate sufficiently for their family maintenance is greatly impeded by the absence of either temperature or rainfall quantity pattern or uniformity. The divergent effects of recent extreme weather events around the world, including within relatively small geographical areas, exemplify the unequal impacts of climate change on populations. Akwa Ibom State has been found vulnerable to extreme weather events, such as flooding, severe storms, and rising sea levels, leading to homelessness, poverty, conflicts, and war for millions of people. All of these have resulted in social disturbances and dislocations among rural populations, especially in coastal communities, making them more vulnerable to climate variability. In the field of social vulnerability in the state, not much has been achieved. This chapter analyzes the vulnerability of the rural population to climate variability; the socio-economic characteristics of the rural population; the index of social vulnerability of rural dwellers to climate variability; social vulnerability factors; and the rural population's social vulnerability mitigation initiatives in Akwa Ibom State, Nigeria. Social science approaches to human vulnerability draw critical attention to the root causes and factors why people are forced to respond to risks from climate change. A complex social approach to vulnerability is most likely to enhance mitigation and adaptation preparation efforts, given that vulnerability is a multidimensional mechanism rather than an invariable state.

## Introduction

Reading through Gallent and Scott's (2017) observations on rural development planning, it gives a great deal of concern and experience that rural planning has been reduced to the least topic of planning theory and practice in more than five decades. A topical issue of rural development planning in the developing countries of the world should be the improvement of the economy at the rural governance level, given that the area is abundantly endowed with human and natural resources. The vast majority of the population still resides in rural areas and depends on the area for food as well as the supply of basic industrial raw materials, regardless of the increased rural-urban migration rate in many developing countries, including Nigeria (UN 2018).

A 2018 United Nations report showed that a quarter of the population in the countries of the Organization for Economic Co-operation and Development (OECD) is mainly rural, while the rural population is 46% worldwide. Critically, the rural economy is characterized by its own people, cultural characteristics, the uniqueness of the natural resource base, social and political institutions, and a combination of rural life which promotes national development. Eighty percent of the population in Nigeria lives in rural areas and their livelihoods are essentially linked to the rural economy (Ering et al. 2014). In addition, the rural area is described as having a low population density, small geographical sizes, and relative isolation, where people are relatively homogeneous in their values, norms, customs, and culture (Ebong 2017). Ejue emphasized that Nigeria's rural areas are characterized by poor and dilapidated roads, inadequate health facilities, inadequate and poorly equipped educational facilities, poor social facilities, high levels of poverty, unplanned residential layout, and other bottlenecks that hinder human well-being.

Notwithstanding these negative aspects, Scott et al. (2019) found that the rural area can be a reservoir of potential such as serene environment, quality land mass, rich biodiversity, tourism entertainment sites, and indigenous local knowledge reservoirs. At the rudimentary level, Frank and Hibbard (2017) also noted that rural areas provide the backdrop for a number of critical planning issues, such as the distribution of renewable energy, the management and improvement of biodiversity services, food production and security, as well as the extraction and administration of natural resources.

Rural people live very closely with nature in rural areas and natural resources provide the basis from which their fundamental needs are derived. Most of the population in rural regions, for example, rely on rain to start farming activities. There is a unique ecosystem for every community in the rural area, valued, admired, and protected by the people living there. For their social, spiritual, and economic development, they rely on those resources. However, due to prevailing climate conditions and associated events such as floods, storms, droughts, and landslides, a good number of rural dwellers are dislocated, especially in the coastal regions (Gallent et al. 2017). For example, some households are known to have moved further inland from their original settlements in Akwa Ibom State, a southern Nigerian state, leading to geographical dislocations (Ebong 2017). In coastal regions, displacement and migration of people from and to different settlements has also been linked to either or both the incidence of drought and accelerated sea-level rise (Abaje et al. 2016). Developing countries are projected to have an uneven share of the impact of climate variability, and due to its geography, climate, vegetation, soils, population and settlement, energy demand, and agricultural activities, Nigeria is particularly vulnerable (Abaje et al. 2016).

Via extremes of weather and interannual fluctuations, climate situations influence both natural and social processes. In some rural areas, with drenching downpour, the capacity of rural households to generate enough to feed themselves is greatly impeded by short rainy seasons (Okpara et al. 2017). These weather patterns influence agricultural production, reduce crop yields, and, in response to the prevailing conditions, enable farmers to change their agricultural practices. In the work of Umoh et al. (2013), it was reported that Akwa Ibom State was vulnerable to adverse impacts of climate variability and change in the Niger Delta region of Nigeria. Considering the widespread poverty, unequal distribution of land, over reliance on rain-fed agriculture, low income, and poor institutional capability, this observation became evident. The State is also vulnerable to extreme weather events such as flooding, severe storms and rising sea levels, in addition to inter-seasonal and inter-annual climate variations, resulting in loss of food crops, household food shortages, homelessness, poverty, and exacerbating circumstances that would lead millions of people to violence, war, and misery (Umoh et al. 2013). In some communities in Akwa Ibom State, heavy and sustained rainfall beyond the absorption capacity of the soil and the flow capacity of rivers and streams is prevalent, and populated areas adjacent to these rivers and streams are subject to recurring flooding, making the region "flood-prone" and placing the population at risk. All of these contribute to social disruption and dislocation of rural populations, especially in coastal communities, making them more vulnerable to climate change.

There is really no consensus that the exact nature of vulnerability renders it a relative concept that is disputed. The word "vulnerability" is derived from a Latin word meaning to be hurt, "vulnerare." This demonstrates the vulnerability of persons and activities to physical violence and/or psychological harm. The vulnerability of a device, individual, or person to a threat is broadly related to the potential to be harmed by that threat (Calderon and Serven 2014). The Intergovernmental Panel on Climate Change (IPCC) describes vulnerability as "the degree to which a system, including climate instability and extremes, is vulnerable to or incapable of dealing with adverse effects of climate change" (IPCC 2014). Vulnerability can have some effect on physical and socio-economic characteristics that are not static. This implies that vulnerability is unique to those evaluating it in a place, context, time, and perspective. Social scientists and climate scientists also use the word "vulnerability" to mean various things, although social scientists prefer to see vulnerability as reflecting the collection of socio-economic factors that decide the capacity of people to cope with stress or change, allowing groups to differ (Allen 2011; Huynh and Stringer 2018). Climate scientists also view a human system's vulnerability as determined by the existence of the physical hazards to which it is exposed, the probability of the hazards occurring regularly and the level of human exposure to hazards, as well as the resilience of the systems to the danger impacts (Ruppert and Deady 2017; Sadri et al. 2017). Vulnerability, therefore, is a combined hazard, exposure, and sensitivity feature (Calderon and Serven 2014).

Most commonly, social vulnerability is described as "the differential ability of groups and individuals to cope with hazards based on their physical and social world roles" (Dow 1992) or as "the inability to take effective loss insurance steps" (Bogard 1989). This also indicates the degree to which the loss, injury, or harm caused by the effects of a climate threat is revealed to a population, physical infrastructure, economic assets, and the general well-being of individuals. Any of these consequences are due to the features of social relationships, organizations and structures, and cultural values. In contrast to the prevailing belief in vulnerability to the impacts of climate variation, which focuses on the physical dimensions, this concept draws more attention to the social dimension of vulnerability. Obviously, social vulnerability focuses on demographic, social, economic, political, and cultural factors that increase the effect of climate hazards on rural inhabitants. In terms of education, gender, income, health status, access to credit and information technology, formal and informal (social) capital, political power, and so on, people living in a group most often differ; these variables also trigger variations in their levels of vulnerability (Huynh and Stringer 2018).

Unfortunately, the social dimensions of vulnerability to climate hazards are rarely paid attention to in most recorded issues and corresponding measures, thus explaining the reason for the very limited studies on social vulnerability in Nigeria taken either at national or state level. In the disaster literature, social insecurity of people in rural areas is widely overlooked and tends to be the least documented condition, explaining why social losses are most frequently missing in estimates of postdisaster cost/loss assessment (White and Howe 2012; UN 2018). Not being able to assess all aspects of climate-related vulnerability would have a detrimental effect on the adaptability of populations and their resilience to climate-related risks. There has been a strong gap in the actual understanding of how socially vulnerable rural communities in Akwa Ibom State emphasize the need to assess their status as socially vulnerable to climate change, taking into account steps to mitigate social vulnerability in rural communities. In order to fill the void, this research came in.

Therefore, this chapter summarizes the report of a state-run study on the social vulnerability of rural dwellers to climate variability. The study described the rural population in the state as specific targets, identified the social vulnerability index of the rural population in the state, identified factors affecting the social vulnerability of the rural population in the state, and identified initiatives to reduce the social vulnerability of the rural population in the state.

## Rural Population Characterization in Akwa Ibom State

Sex, marital status, age, educational qualification, primary occupation, monthly income, household size, form of housing, position of building, and health status were the selected socio-economic characteristics of the respondents considered in the sample. The respondents' sex distribution showed that there were more females in the areas than males. The dominance of women in the fields of study can be attributed to the migration of men to the source of alternative opportunities for urban jobs. It also means that with climate change, more women are likely to be affected because in the study region they outnumbered the men. During rehabilitation, women will have more difficult times than men, due to streamlined jobs, meager salaries, and responsibility for family care. In local and national decision-making, women often lack a voice and a representative and are isolated in political processes, affecting their exposure, sensitivity, and resistance to climate hazards (Mansur et al. 2016; Anderson 2014).

Women account for the greater majority of the world's poor in rural areas. They depend on natural resources for their livelihoods and survival more than men. Women tend to have lower earnings and are more likely than men to be economically dependent. For example, when droughts or unseasonable rains threaten agricultural production, men may use their savings and economic freedom to invest in or otherwise adapt to alternative sources of income (Omi and Winant 2015). But as household shock absorbers during periods of economic and physical deprivation, women extend their already highly undervalued working hours and may even decrease their nutritional status to increase that of the family (Maldonado et al. 2016).

The majority of respondents were married during the research. The high percentage of married respondents is consistent with Ekanem et al. (2020), who noted that among rural people in Nigeria, getting married is highly valued with the intention of using these women for unpaid family work. The high percentage of "married" marital status appears to impose certain restrictions on most female study respondents because married women are forced to adhere to certain requirements imposed by their husbands directly or indirectly by tradition. Men's dominant efforts make women feel inferior and unable to take part in life-affecting events. For example, a female respondent said, "When disaster occurs, without the consent of our husbands, we cannot move."

In the research area, the study found a certain percentage of widowed respondents. This may be attributed to a high mortality rate among men drowning in the high seas during an expedition to fish in the coastal parts of the state. During the analysis, the percentage of the separated according to the extracts from the FGD groups is due to the loss of housing during the flood time that makes couples absolutely powerless as they have to transfer to nonaffected areas. One of the women had to say, "Our husbands migrate from rural areas to cities to exploit perceived economic opportunities in the course of economic deprivation as migrants and indirectly dumb us at home to take care of children, elderly or sick family members and often remarry other women outside our culture."

The bulk of the respondents were between the ages of 61–80. Because of their health problems, those respondents who were within the age range of 60–80 years were considered to be economically inactive, rendering them unable to embrace new information or to engage in adaptation preparation in order to minimize their degree of social vulnerability to climate change. The educational level of rural residents can restrict their ability to understand and respond to climate information. The findings of our study showed that the majority of respondents (64.7%) did not have formal education, 23.7% had primary education, 11.0% had secondary education, while just 0.67% of respondents read up to tertiary level. This means that the majority (64.7%) of respondents would be more vulnerable to climate change due to their inability to access, read, and comprehend weather information literature, as posited by Davies et al. (2017), who noted that various climate threat information cannot be accessed or acted upon by the least informed community members in order to plan for recovery.

The majority of respondents had agriculture as their primary occupation, with few others taking on employment in commerce, crafts, and wages. Other occupations, such as boat mending, water transport, motorcycling, etc., constituted 1.7% of the primary occupation of the respondents. The outcome showed that 40.0% of the respondents were artisans engaged in various low-income activities. During catastrophes, this behavior is typically lost and therefore increases their vulnerability and resilience.

Most respondents had a monthly income of N1,000–20,000, with just 1% of respondents receiving a monthly income of N40,001–N60,000. The outcome indicates that 95% of low-income (1000–20,000) respondents had little to no financial support to mitigate their social vulnerability. Ekanem et al. (2020) and Anderson (2014) expressed the view that households with less access to energy are more economically sensitive to the impacts of immediate and eventual climate variability.

Household sizes are very high for rural dwellers in Akwa Ibom State. Our study found that most respondents had household sizes in the range of 11–15 individuals. Due to frequent flooding that damages residential apartments causing population displacement, it can be deduced that household size increases, which could result in more individuals sharing few habitable apartments. Some respondents, especially the coastal residents, pointed out that "we lost our houses when disaster hits and plead with neighbors to squat us." The standard of the housing is closely linked to the personal wealth of the respondent. That is why some rural residents live in badly developed houses that are vulnerable to heavy rainstorms and flooding (Hollesen et al. 2018). Our analysis has revealed 79. Nine percent of the respondents lived in thatched roof mud buildings, while 8.3%, 5.7%, and 6.3% of the respondents lived in zinc roof, zinc mud, and zinc cement, respectively. As Cutter et al. (2016) and Kintisch (2016) asserted that rural poor live in substandard and overcrowded houses vulnerable to the spread of diseases and the effect of disasters, the results indicate that respondents are mostly poor. Of necessity, this is a vital element in assessing the vulnerability of climate hazards. In rural areas, the value, quality, and density of residential buildings affect possible losses and recovery.

Observations have shown that, as demonstrated by the 2009 U.S. Hurricanes Katrina, climate threat and its aftermath have a significant effect on people with disabilities. Our analysis found that 19.9% of the participants were people with disabilities. These individuals are likely to be more vulnerable to the consequences of climate variability as "climate change causes increasing distress and reduced opportunities for them to access critical emergency information" (Cutter et al. 2016; Hollesen et al. 2018). The World Report on Disability also reveals that people with disabilities are particularly inclined to be neglected in emergency assistance and relief operations (World Health Organization and World Bank 2011) (Table 1).

## Social Vulnerability of Rural Dwellers to Climate Variation in Akwa Ibom State

Table 2 indicates many factors that predispose households to the adverse effects of the state's climate variability. Ten elevated rank variables have been picked. The first three reasons were: dwelling place loss ($x = 2.38$; ranked 1), properties open to criminals ($x = 2.32$; ranked 2), and children's failure to complete their education ($x = 2.29$; ranked 3). The results suggest that the nature and location of housing makes individuals more vulnerable to risks. In hazard-prone areas, such as flood plains and slums, poor people are more likely to live. This

**Table 1** Distribution of rural dwellers based on their socio-economic characteristics

| Item | Selected variables | Frequency n = 300 | Percentage (%) |
|---|---|---|---|
| 1 | **Sex** | | |
| | Male | 142 | 47.3 |
| | Female | 158 | 52.7 |
| 2 | **Marital status** | | |
| | Single | 33 | 11.0 |
| | Married | 150 | 50.0 |
| | Separated | 27 | 9.0 |
| | Widowed | 90 | 30.0 |
| 3 | **Age (years)** | | |
| | <1–20 | 1 | 0.3 |
| | 21–40 | 3 | 1.0 |
| | 41–60 | 50 | 18.7 |
| | 60–80 | 240 | 80.0 |
| 4 | **Educational status (years of formal schooling)** | | |
| | 0 | 194 | 64.7 |
| | 1–4 | 71 | 23.7 |
| | 5–8 | 33 | 11.0 |
| | 9–12 | 2 | 0.67 |
| 5 | **Primary occupation** | | |
| | Farming | 66 | 22.0 |
| | Trading | 88 | 29.3 |
| | Artisan | 120 | 40.0 |
| | Salaried job | 21 | 7.0 |
| | Others (Ukada) | 5 | 1.7 |
| 6 | **Income level (N)** | | |
| | <1–20,000 | 289 | 95.3 |
| | 20,001–40,000 | 9 | 3.0 |
| | 40,001–60,000 | 4 | 1.3 |
| | 60,001–80,000 | 1 | 0.3 |
| 7 | **Household size** | | |
| | 1–5 | 27 | 9.0 |
| | 6–10 | 115 | 38.3 |
| | 11–15 | 158 | 52.7 |
| 8 | **Housing type** | | |
| | Thatch roof | 239 | 79.7 |
| | Zinc roof | 25 | 8.3 |
| | Mud with zinc | 17 | 6.3 |
| | Cement with zinc | 19 | 6.3 |
| 9 | **Location of building** | | |
| | Low lying land | 15 | 5.0 |
| | Flood plain | 19 | 6.3 |
| | Close to seashore | 56 | 18.7 |
| | Upland | 210 | 70.0 |

(continued)

**Table 1** (continued)

| Item | Selected variables | Frequency n = 300 | Percentage (%) |
|---|---|---|---|
| **10** | **Health status** | | |
| | Not physically challenged | 240 | 80.0 |
| | Hearing/speech impaired | 12 | 4.0 |
| | Visually impaired | 19 | 6.3 |
| | Amputated | 8 | 2.6 |
| | Hypertensive/diabetic patient | 21 | 7.0 |

corroborates the earlier findings of Birkmann (2013) and Nelson et al. (2016), who recorded that hastily established settlements are more physically vulnerable in terms of subsistence and geographical position to immediate climate impacts. Nelson et al. (2016) further clarified that rural communities have less political voice in disasters such as severe flooding and windstorms, thus raising the likelihood of their residents being physically harmed. Brown (2016) argued that the most vulnerable to climate hazards are likely to be the least trained and low-skilled members of society and that government may pay little attention to their well-being. This implies, as Tall (2019) notes, that vulnerability is reduced by better education and knowledge. Suitable weather knowledge, for example, enables farmers to predict seasonal trends.

Variables in the 4th, 5th, and 6th ranks were loss of privacy (x = 2.27), blocked from health facilities (x = 2.20), local workshop destruction (x = 2.18). During floods, the privacy of people tends to be absolutely threatened, eventually subjecting them to immense social deprivation. The women had this to say in the study area (coastal region); in many cases, due to lack of privacy, we cannot respond to the call of nature. As a consequence, to defecate, we have to wait till night. In the case of pregnant mothers, disabled, and elderly people, the condition is getting worse. The study by Brondızio et al. (2016) and Hollesen et al. (2018) reinforced that during floods, disease outbreaks are prevalent; they submerge latrines and contribute to misery. They not only become victims of snake bites, but also suffer from skin diseases regularly. People in poor health and those who are undernourished will be more vulnerable to the immediate and secondary effects of extreme climate events, according to Hardy et al. (2017), since they are less likely to be able to plan for or cope with the impacts of climate hazards.

The ranked variables placed 7th, 8th, 9th, and 10th were: absence of services (x = 2.16), evaluation of relief materials (x = 2.15), loss of relatives (x = 2.11), and lack of drinking water (x = 2.10). These results have reinforced the assertions of Gamble et al. (2010), who identified the need to take into account the role of human and social capital in assessing climate impact vulnerability. They believed that human and social resources can contribute to the capacity of a society to overcome climate variability through coping and response capacities, whereas a lack of capital can leave individuals isolated and at higher risk of impacts such as severe heat waves being revealed.

**Table 2** Social vulnerability of rural dwellers to climate variability

| S/N | Statements | SD | D | A | SA | Mean | Rank |
|---|---|---|---|---|---|---|---|
| 1 | We hardly see our relatives as they have been dislocated because of constant flood | 33(11.0) | 72(24.0) | 103(34.3) | 92 (30.7) | 2.1 | 9 |
| 2 | Our local workshop has been destroyed by constant flooding | 18(6.0) | 110(36.7) | 111(37.0) | 61(20.3) | 2.18 | 6 |
| 3 | Peoples' houses are pulled down by strong wind in my community | 5(1.7) | 99(33.0) | 108(3.0) | 88(29.3) | 1.90 | 15 |
| 4 | There is no safe drinking waters in my community as a result of flood | 44(14.7) | 48(38(12.7) | 113(37 3) | 95(31.7) | 2.10 | 10 |
| 5 | People lost their homes due to severe storm | 5(1.7) | 38(12.7) | 94(31.3) | 163(54.3) | 2.38 | 1 |
| 6 | Our food crop is lost because of rise in water level | 6(2.7) | 49(16.3) | 128(42.0) | 117(39.0) | 1.72 | 19 |
| 7 | Our children cannot complete their education because school facilities are flooded | 16(5.3) | 30(10.0) | 105(35.0) | 149(49.7) | 2.29 | 3 |
| 8 | People are attack with different diseases such as skin infection, malaria, typhoid, and dysentery because of continuous inflow of water | 70(23.3) | 32(10.7) | 69(23.0) | 129(43.0) | 1.86 | 16 |

(continued)

**Table 2** (continued)

| S/N | Statements | SD | D | A | SA | Mean | Rank |
|---|---|---|---|---|---|---|---|
| 9 | Our community is blocked from healthcare services because of increase in water level | 10(3.3) | 41(13.7) | 128(42.7) | 121(40.3) | 2.20 | 5 |
| 10 | We cannot attend ceremonies like wedding child dedication in nearby communities due to high level of water | 77(25.7) | 32(10.7) | 66(22.0) | 125(41.7) | 1.80 | 17 |
| 11 | Our properties are open to thieves because of lack of accommodation | 85(28.3) | 29(9.7) | 55(18.3) | 131(43.7) | 2.32 | 2 |
| 12 | We do not have privacy where we are squatting | 52(17.3) | 97(32.3) | 48(160) | 154(51.3) | 2.27 | 4 |
| 13 | Our traditional food and dialect are disappearing as we move to new communities. | 21(7.0) | 30(10.0) | 44(14.7) | 112(37.3) | 1.79 | 18 |
| 14 | Our children get involve in hard jobs to contribute to family's income | 31(10.3) | 113(37.7) | 36(12.0) | 171(57.0) | 2.03 | 11 |
| 15 | People with disabilities have lost assistance and care from relatives as flood disrupt their dwelling places | 29(9.) | 64(21.3) | 69(23.0) | 155(51.7) | 2.18 | 6 |

(continued)

**Table 2** (continued)

| S/N | Statements | SD | D | A | SA | Mean | Rank |
|---|---|---|---|---|---|---|---|
| 16 | There is reduction in our livestock production as floods pull down farmsteads | 25(8.3) | 51(17.0) | 89(29.0) | 136(45.3) | 1.71 | 20 |
| 17 | We cannot recover our lost properties | 28(9.3) | 47(15.7) | 111(37.0) | 98(32.7) | 1.93 | 14 |
| 18 | We lack facilities to reach out to our customers to sell our farm produce | 29(9.7) | 62(20.7) | 122(40.0) | 106(35.3) | 2.16 | 7 |
| 19 | We are cut off from neighboring community | 26(8.7) | 46(15.3) | 109(36.0) | 124(41.3) | 1.99 | 13 |
| 20 | We cannot assess relief materials in my community | 26(8.7) | 41(13.7) | 99(33.0) | 120(40.0) | 2.15 | 8 |
| 21 | There is scarcity of hired or paid labor in our community due to migration | 36(12.0) | 45(15.0) | 66(22.0) | 155(51.7) | 2.01 | 12 |
| 22 | Our traditional rulers do not inform government about our situation | 31(10.3) | 48(16.0) | 97(32.3) | 103(34.3) | 1.67 | 21 |

KEY
SD = Strongly Disagree
D = Disagree
A = Agree
SA = Strongly Agree

Rural dwellers seem vulnerable because they cannot access the services of their society to cope with climate change shocks and stresses. Hardy et al. (2017) argued that their choices are limited and very difficult for them to think beyond their immediate needs, let alone make adaptation plans, if people do not have secured access to vital livelihood tools. Climate change disrupts social relationships, undermining family bonds, and social ties with loved ones.

## Factors Affecting Social Vulnerability to Climate Variation

Factor analysis was used to examine and identify the factors affecting social vulnerability to climate variation. Sixty (60) items were initially generated during the instrument construction phase to reflect the objective after which it was subjected to factor analysis using principal component approach. This was done to reduce the number of items to major and sizeable significant number of factors using Eigen value criterion of $\geq 1$ to select the underlying dimensions of the original 60 items.

The principal components method is an exact mathematical transformation of the original set of variables to a new set, with the later summarizing linear relationships exhibited in the data. The best linear combination of the variables gives the first component or factor of the variables. It means that this factor, more than any other, accounts for most of the variance or relationships observed in the data and hence is the best summary of the original data. The second component is similarly defined as the next best linear combination of the variables that would account for the relationships observed in the data after the effect of the first component or factor has been removed, provided both components are uncorrelated (orthogonal). Subsequent components are similarly defined until all the relationships or variance in the data have been accounted for. The factor can therefore be defined as:

$$F = aWA + bwB \pm \ldots kwK \qquad (1)$$

where

Fa = factor a, being the best linear combination of all the variable A-K
aw-kw = are weights attached to variables
A-K = variables A-K

Other factors can be similarly defined, as explained earlier. When the factors are uncorrelated, they are said to be orthogonal and a variable "a" in the original data can be defined as:

$$Z_a = Wa_1F_1 + W_{a2}F_2 + \ldots + W_{a2}F_{2+} \ldots + W_{ak}F_k \qquad (2)$$

where $Z_a$ = Variable a in standard score format (mean = 0 and variance 1), $W_{a1}$ to $W_{ak}$ = Weights (normally called factor loadings) attached to factors, and $F_1$—$F_k$ = factors I–K. Formula 2 simply means that in factor analysis employing the principal components methods, a variable is viewed as the sum of weighted factors derived from the data. When the data is initially factored, the concern is more on the possibility of reducing the data to a smaller composite set of factors. At this stage, the factors may not make any meaning until they are rotated to a final solution.

## Empirically Defining the Underlying Dimensions of Factors Affecting Social Vulnerability

Seventeen (17) composite factors were mutually exclusive on the basis of the result of factor analysis performed and major trends were produced. Table 3 introduces these variables and addresses them below:

### Local Support Factor

Inability to access community services is an important aspect of social insecurity, as it relates to the way rural dwellers are categorized on the basis of their income, according to literature (Birkmann 2012). Factor 1 accounts for 21.7% of variance and this is high. This outcome coincides with Passel's 2005 study that poorer households (low class) are not able to access resources fairly due to social factors (economic and political processes) that are custodians of upper class or wealthier people.

**Table 3** Major dimensions of social vulnerability to climate variability

| Factor | Names | Eigen values | % of variance | Cumulative % |
|---|---|---|---|---|
| 1 | Lack of local support | 12.995 | 21.658 | 21.658 |
| 2 | Knowledge barrier | 3.762 | 6.270 | 27.928 |
| 3 | Income/employment/replacement ability | 2.581 | 4.302 | 32.230 |
| 4 | Family structure | 2.333 | 3.888 | 36.118 |
| 5 | Demographic explosion | 2.129 | 3.548 | 39.667 |
| 6 | Age of residents | 1.851 | 3.085 | 42.751 |
| 7 | Gender inequality | 1.710 | 2.849 | 45.601 |
| 8 | Tribal issue/public assault and insecurity | 1.564 | 2.607 | 48.208 |
| 9 | Infrastructure inaccessibility | 1.542 | 2.571 | 50.779 |
| 10 | Loss of properties and accommodation | 1.494 | 2.490 | 53.268 |
| 11 | Mobility | 1.407 | 2.346 | 55.614 |
| 12 | Disability | 1.291 | 2.152 | 57.766 |
| 13 | Weak local institution | 1.225 | 2.041 | 59.808 |
| 14 | Cultural constraint | 1.194 | 1.991 | 61.798 |
| 15 | Poor climate information/communication source | 1.173 | 1.955 | 63.753 |
| 16 | Loosed social capital/social safety net | 1.070 | 1.783 | 65.535 |
| 17 | Environmental/social discrimination | 1.051 | 1.752 | 67.287 |

## Educational Challenge Factor

Training is related to socio-economic status and accounts for 6.2% of rural dwellers' heterogeneity. Noneducational achievement in the field of study makes community members vulnerable to climate danger because they lack the opportunity to grasp knowledge on warning and recovery. The outcome of the finding is consistent with Nelson et al. (2016), who stated that during floods, children's schooling is impaired as it is difficult to keep and attend classes under submerged conditions. The school appears to become the "front-line" flood shelter for displaced families anywhere the school building is not submerged. UNICEF (2011) confirms this finding by stating that children are forced out of school when natural disasters interrupt the educational system of the community, including teachers. The absence of experienced teachers and other main school staff is detrimental to the academic success of children.

## Income /Employment/Replacement Ability Factor

Floods have a significant effect on the source of livelihoods, paralyze economic activities, deepen disparities in economic status, and access to sufficient food and services in the field of research. This explains the 4.3% difference between rural communities. The result supports Sultana (2017) who reported that when affected by a danger case, local individuals who are in primary extraction activities such as fishing, forestry, crop cultivation, and livestock rearing suffer great losses and are unable to rebuild their settlement.

## Family Structure Factor

When tragedy occurs, large family size hinders evacuation and this impact increases social vulnerability. There is less time, resources, and energy for families caring for ill members to invest in activities that could minimize the impacts of external threats, help them recover from dangerous incidents, and be better prepared for them in the future. Sultana (2017) also acknowledges that members of large and dependent families – children under the age of 18, elderly and disabled members – need financial assistance, transportation, and disaster medical treatment, and these affect the resilience to risk recovery. This portion explains the variability of 3.8% among respondents.

## Demographic Explosion Factor

Village habitation is adversely impacted by population migration and neighborhood disorder. When the flood effect of the community or household shifts to the non-affected area, there is a population influx leading to an increased risk of disease and sanitation system contamination. This is corroborated by McLeman (2016), who stressed that high population increases outbreaks of disease due to individual proximity to each other and contamination of human waste water sources. Increasing population density increases vulnerability, particularly where the density is high, according to local respondents. Increased population density limited access to land and exacerbated pressure on natural resources during challenging times, respondents clarified. Indeed, population growth means that growing numbers of people are exposed to the impacts of climate variations and that more people have to share scarce resources (e.g., water and food). This explains the 3.5% difference between rural dwellers.

## Age of Resident Factor

Age is strongly associated to social disadvantage and 3.0% of the difference is clarified. There is also a rise in health challenges as individuals grow older, and these are compounded during times of dislocation or loss of livelihood due to climate change. The result confirms Omi and Winant (2015), who reported that due to adverse effects of climate variability such as hearing impairment, cognitive difficulties, rise in blood pressure (hypertension), etc., the elderly population experiences different health problems. This affects their potential for disaster management and recovery. Brondızio et al. (2016) also argued that children cannot protect themselves during a disaster, particularly the youngest age group, because they lack the resources, skills, or life experiences required to cope effectively with the situation.

### Gender Inequality Factor

2.8 percent of the observed variance is explained by this factor. Sex is a powerful factor illustrating the susceptibility to impacts related to climate change. Factors such as lack of access to and power over basic resources and lack of entitlements exacerbate the vulnerability of women and weaken their capacity to cope with disaster effects (Sultana 2014). The result is consistent with Brown (2016) which stated that because of their distinct socially constructed positions as mothers and family caregivers, women suffer more from the impact of disaster. Faber (2015) notes that stream dries up during extreme situations, for example, when natural water sources, and women have to travel longer distances daily on foot in search of drinking water. The report shows that pregnant and lactating mothers make up a large proportion of these women. Furthermore, he added that women enumerated their experiences during constant flooding over menstrual hygiene management, especially in the coastal zone, resulting in genital injury, bleeding, infections, and other complications.

### Tribal Issues, Public Assault and Insecurity Factor

2.6 percent of the variance among respondents is explained by this factor. Literature indicates that people who lose their homes during floods are absolutely powerless as they travel out to non-affected areas without any source of income. They lose the physical social security they once enjoyed in their native villages in their new housing as migrants. This, in turn, puts them in different kinds of adverse social circumstances; young girls in constant danger of sexual abuse and attack from the boys and men of that group who take shelter on embankments. Professional gangs are also coerced into prostitution by young women, with offers of work elsewhere (Maldonado et al. 2016). The outcome also agrees with Dunning (2011), who noted that, for example, people who do not speak the language of their host group have trouble understanding and reacting to the warning and evacuation order

#### Infrastructure Inaccessibility Factor

2.5% of the variance is explained by this factor. The outcome is consistent with Davies et al. (2017), who stressed that social deprivation is increased by inaccessibility to vital infrastructure such as highways, shelter, and telecommunication. He emphasized that poorly maintained local road networks rendered some roads impassable. The other part of the city, including the economy, is cut off and blocked from some areas that would promote income security.

#### Loss of Properties/ Accommodation Factor

This factor explains 2.4% of the difference in rural populations. The outcome is in line with Brondızio et al. (2016) who highlighted the flood "sweep off" farmland leading to crop and income loss. In particular, thatch houses are said to be damaged in coastal communities by floods in the upland or sea level buildings. He further claimed that, especially for poor households, loss of property is more difficult to replace. Some fishermen describe their bitter experiences during a sea pirate attack fishing trip on the high seas where their boat engines and other instruments are stolen and destroyed.

#### Mobility Factor

This portion explains the variation among respondents of 2.3%. The outcome of the results agrees with Cutter et al. (2016) who posited that for people who do not have cars, transport out of the evacuation zone is problematic. In coastal regions, the movement of people is distorted by the increase in sea level because canoes made locally cannot be paddled at high sea level, are victims of water-related diseases, and cannot access medical facilities and workers in upland areas leading to trauma of nonpathogenic hypertension and stroke conditions. Brown (2016) claims that when emergency strikes, people with disabilities have mobility restrictions, they have trouble finding safe places.

#### Disability Factor

This variable explains the difference of 2.2% among rural dwellers. The finding is synonymous with the World Disability Study (2011) that when emergency strikes, people with disabilities are left behind in a deteriorated community and are isolated from their families and friends, leading to crippling suffering on them, and they lack the social networks needed to take advantage of services that would speed up their disaster recovery. Hollesen et al. (2018) reported that disasters and their effects have a significant effect on people with disabilities due to the lack of assistive devices such as wheelchairs, crutches, white canes, or hearing aids.

### Weak Local Institution Factor

2.0% of the differences among local dwellers are explained by this factor. Poor or corrupt institutions can contribute to ineffective and insufficient responses to disaster situations, slowing requests for government support and not actually delivering help to those most in need. This supports Ejue, who claimed that rural communities are limited by weak leadership and lack of freedom and decision-making power as their voices, needs, and desires are not integrated into government projects and this increases sensitivity for poorer households.

### Cultural Constraints Factor

This element explains the variance of 1.99%. The result is the same with Ejue, who argued that the lack of traditional wisdom has caused elders to suffer contempt from the younger generation because the wisdom they carry is not as detailed as it used to be. This outcome also agrees with Mansur et al. (2016), who stated that as their ethnic cultural identity suffers from the risks of extinction as a result of climate-related risks, older people become more dissatisfied.

### Climate Information/ Communication Factor

The lack of available, affordable, and appropriate sources of knowledge leads to social vulnerability. This explains 1.95% of the variance. It was found that respondents rely on their personal weather observation, past experience in their community and local institution due to irregularities, and untimely distribution of advance media information that they caption "false stories" or "fly news." Their conviction is due to the fact that localities are not subject to weather predictions from radio and NTA stations. This result is consistent with d'Alpoim et al. (2016), which posits that in terms of reliability, timing, and languages, there are shortcomings in the knowledge distribution process. It is important to communicate, obtain, and interpret information about weather forecasts and climate predictions via multiple channels, including interpersonal contact, television, radio, the Internet, and social media (Morss et al. 2017). Before they make decisions, individuals combine data from these different sources (Sadri et al. 2017). Social networks help people access, personalize, and understand the importance of knowledge in a significant way.

### Loosed Social Network and Safety Net Factor

1.78 percent of the variance is explained by this factor. The study agrees with Cutter et al. (2016) that it is difficult for displaced populations to re-organize and restore their former family connections and relationships. Some respondents share their feelings that floods are hindering social activities such as marriage, commitment to children, and funerals. This condition weakens social capital and loosens social relations with loved ones, causing the respondents to be mentally depressed. Kintisch (2016) argued that the loss of the network and lack of access to social gathering make people feel lonely, especially the elderly. Meanwhile, Davies et al. (2017) assume that if placed in place, the social security net will help prevent postdisaster starvation and assist affected households and communities to safeguard and restore their properties.

## Environmental / Social Discrimination Factor

This element explains the variance of 1.75%. The outcome agrees with Cutter et al. (2016) and McDowell et al. (2016), who observed that a nation's inhabitants do not share the benefits of economic growth equally. Discrimination and social isolation result in this. And in rural areas, this is also true.

## Measures to Reduce Rural Residents' Social Vulnerability to Climate Change

Table 3 demonstrates strategies that will reduce the social vulnerability of respondents in the study region to climate change. Ten of the most relevant high-rank indicators were chosen. The first four steps to mitigate social vulnerability to the effects of climate variation were: dam construction (x = 3.62; ranked 1), flood storage reservoir construction (x = 3.57; ranked 2), dam and bridge construction (x = 3.56; ranked 3), and gutter opening (x = 3.54; ranked 4). These initiatives agree with Eriksen (2012), who noted that one of the adaptation techniques widely practiced by flood victims is the construction of defensive outlets such as flood storage reservoirs, dams, and water transport channels from the affected areas. Davies et al. (2017) accept that for uprooted communities, the embankment acts as temporary shelter.

The fifth method is working to fill sand bags in groups (x = 3.53, ranked 5). This agrees with McDowell et al. (2016), who stated that for mutual benefit, societies with denser social support networks have more interaction between community members. Eriksen (2012) believed that societies with higher social and stronger social networks will be better able to cope with threats and their consequences because it is easier to analyze knowledge and other types of social support.

The sixth technique is the fund's payment (x = 3.52, ranked 6). The results confirm that Huynh and Stringer (2018) emphasized that materials for relief and assistance are especially helpful to those individuals who could not afford to repair damaged properties themselves. The seventh, eighth, ninth, and tenth steps were: environmental law enforcement (x = 3.50, ranked 7), flood-prone areas should not be developed (x = 3.49; ranked 8), weather signal recognition (x = 3.47; ranked 9). The result corroborates the view of Omi and Winant (2015), who said that individuals should be prevented from developing areas vulnerable to flooding and that defaulters should be punished. And the use of local dialects (x = 3.37; rank 10). The results indicate that the combination of extension agents, radio, and local agencies will be more effective in disseminating local dialect environment information and supporting services in insecure areas for vulnerable people.

The least significant steps were: raising consciousness of income diversification (x = 3.34; ranked 11), building a well-equipped hospital (x = 3.30; ranked 12), the proportion of local cultivars (x = 3.02 ranked 13), and the lack of equitable access to local services (x = 2.96; ranked 14).

## Conclusion

It can be inferred from the results obtained that 52% of women were vulnerable to climate change in rural areas of Akwa Ibom State. Most of the respondents were between 60 and 80 years of age. The majority of respondents (64%) did not have formal education and are probably the least educated on climate issues. The most important steps to mitigate flooding from the perspective of the respondents were the construction of embankments, flood storage reservoirs, dams and bridges, and opening of gutters; and these were related to the position of respondent construction. The underlying dimensions affecting social vulnerability were lack of local support, information barrier, jobs, employment and replacement capability, family structure, demographic explosion, resident age factor, gender, tribal problems, public attack, and inaccessibility of infrastructure in the insecurity factor. Others included property and housing damages, mobility, disability, weak local institutions, cultural constraints; inadequate sources of environment knowledge and communication; loosened social networks and social security networks; and social discrimination. Rural residents in Akwa Ibom State were highly vulnerable to the effects of climate change because, as seen in the report, the higher the factors that lead to social vulnerability, the higher the degree of social vulnerability of the respondents.

## Recommendation

The following recommendations are made based on the study's results and conclusions:

1. Agricultural extension staff, meteorological offices, and community-based agents should be assisted by policy makers and development partners to raise awareness of climate information forecasting and dissemination through usable and accessible communication methods such as local dialect radio, city critics, drama, newsletters, community workshops/seminars, and educational programs
2. Income diversification is an efficient way to reduce vulnerability by spreading risks. To minimize social vulnerability to extreme flooding and drought, less reliance on natural resources would help. Rural dwellers need to maintain and engage effectively in those value-added economic activities that will reduce food insecurity, increase jobs and income opportunities, and enable individuals to accumulate assets that will enhance their ability to cope with potential shocks without worsening poverty.
3. Adaptive planning is required to minimize the vulnerability of vulnerable coastal roads (ferry routes), low-lying buildings and assets, transport, and market connections and to improve legislation to safeguard hazard-prone areas. Adaptive planning is needed.
4. Local institutions should improve social capital, that is, family ties and friendly connections between members of the group should be revitalized so that individuals can collaborate and team up to resolve the effect and challenges of climate

variations for mutual benefit among themselves. In the meantime, friendly ties with relatives away from home should be established for remittances to act as a cushion when subjected to extreme events and also for the provision of temporary accommodation during displacement.

## References

Abaje IB, Sawa BA, Iguisi EO, Ibrahim AA (2016) Impacts of climate change and adaptation strategies in rural communities of Kaduna State, Nigeria. Ethiop J Environ Stud Manag 9(1): 97–108

Allen R (2011) Vulnerability reduction and the community-based approach. In: Pelling (ed) Natural disasters and development in a globaliing world. Routledge, London, pp 170–118

Anderson EN (2014) Caring for place: Ecology, ideology, and emotion in traditional landscape management. Walnut Creek, CA: Left Coast Press

Birkmann J (2012) Vulnerability reduction: a task for the vulnerable people themselves. In Bank off, et al (eds) Mapping vulnerability: disasters, development and Peter. Earthscan, Sterling

Birkmann J (2013) Measuring vulnerability to natural hazards: towards disaster resilient societies. United Nations University Press, Bonn

Bogard WC (1989) Bringing social theory to hazards research: conditions and consequences of the mitigation of environmental hazards. Sociol Perspect 31:147–168

Brondızio ES, de Lima AC, Schramski S, Adams C (2016) Social and health dimensions of climate change in the Amazon. Ann Hum Biol 43:405–414

Brown K (2016) Resilience, development and global change. Routledge, London/New York

Calderon C, Serven L (2014) Infrastructure, growth, and inequality: an overview. World Bank policy research working paper. The World Bank, Washington, DC

Cutter SL, Ash KD, Emrich CT (2016) Urban– rural differences in disaster resilience. Ann Am Assoc Geogr 106:1236–1252

d'Alpoim Guedes JA, Crabtree SA, Bocinsky RK, Kohler TA (2016) Twenty-first century approaches to ancient problems: climate and society. Proc Natl Acad Sci 113(51):14483–14491

Davies G, Frausin G, Parry L (2017) Are there food deserts in rainforest cities? Ann Am Assoc Geogr Adv. https://doi.org/10.1080/24694452.2016.1271307

Dow K (1992) Exploring differences in our common future(s): the meaning of vulnerability to global environmental change. Geoforum 23:417–436

Dunning MC (2011) Social vulnerability analysis methods for crops planning. Draft report 10/29/09

Ebong M (2017) Mobilisation of resources for rural development in Nigeria. Wusen Press, Calabar

Ekanem JT, Nwachukwu I, Umoh IM (2020) Post-harvest Adaptiveness of crop farmers to future climates in the Cross River floodplains, Nigeria. J Commun Commun Res 5(1):50

Eriksen SH (2012) Responding to global changes vulnerability and management of local agro-ecosystems in Kenya and Tanzania. University of East Anglia, Norwich

Ering SO, Otu JE, Archibong EP (2014) Rural Development Policies in Nigeria: A Critical Appraisal. Int J Educ Res 2(9):307–320

Faber JW (2015) Superstorm Sandy and the demographics of flood risk in New York City. Hum Ecol 43(3):363–378

Frank KI, Hibbard M (2017) Rural planning in the twenty-first century: context appropriate practices in a connected world. J Plan Educ Res 37(3):299–308

Gallent N, Scott M (eds) (2017) Rural planning and development. Routledge, London

Gallent N, Tewdwr-Jones M, Hamiduddin I (2017) A century of rural planning in England: a story of fragmentation, contestation and integration. Planum J Urban 35(2):91–104

Gamble DW, Campbell D, Allen TL, Barker D, Curtis S, Mcgregor D, Popke J (2010) Climate change, drought, and Jamaican agriculture: Local knowledge and the climate record. Annals of the Association of American Geographers 100:880–893. https://doi.org/10.1080/00045608.2010.497122

Hardy RD, Milligan RA, Heynen N (2017) Racial coastal formation: the environmental injustice of colorblind adaptation planning for sea-level rise. Geoforum 87:62–72

Hollesen J, Callanan M, Dawson T, Fenger-Nielsen R, Friesen T, Jensen A, . . . Rockman M (2018) Climate change and the deteriorating archaeological and environmental archives of the Arctic. Antiquity 92(363):573–586. https://doi.org/10.15184/aqy.2018.8

Huynh LTM, Stringer LC (2018) Multi-scale assessment of social vulnerability to climate change: an empirical study in coastal Vietnam. Clim Risk Manag 20(2018):165–180

IPCC (2014) Summary for policymakers. In: Field CB, Barros VR, Dokken DJ, Mach KJ, Mastrandrea MD, Bilir TE, White LL (eds) Climate change 2014: impacts, adaptation, and vulnerability. Contribution of working group II to the fifth assessment report of the Intergovernmental Panel on Climate Change. Cambridge University Press, Cambridge, UK/New York, pp 1–32

Kintisch E (2016) History is melting. Hakai Magazine. Retrieved from http://bit.ly/1PSoFxb

Maldonado J, Lazrus H, Bennett S-K, Chief K, Dhillon CM, Gough B, Whyte KP (2016) The story of rising voices: facilitating collaboration between indigenous and Western ways of knowing. In: Companion M, Chaiken M (eds) Responses to disasters and climate change: understanding vulnerability, building resilience. CRC Press/Taylor & Francis Group, Boca Raton, pp 15–26

Mansur AV, Brondızio ES, Roy S, Hetrick S, Vogt ND, Newton A (2016) An assessment of urban vulnerability in the Amazon Delta and estuary: a multi-criterion index of flood exposure, socio-economic conditions and infrastructure. Sustain Sci 11(4):625–643

McDowell G, Ford J, Jones J (2016) Community-level climate change vulnerability research: trends, progress, and future directions. Environ Res Lett 11(3):33001

McLeman R (2016) Impacts of population change on vulnerability and the capacity to adapt to climate change and variability: a typology based on lessons from "a hard country". Popul Environ 31:286–316

Morss R, Demuth J, Lazrus H, Palen L, Barton C, Davis C, . . . Watts J (2017) Hazardous weather prediction and communication in the modern information environment. Bull Am Meteorol Soc 98(12):2653–2674

Nelson MC, Ingram SE, Dugmore AJ, Streeter R, Peeples MA, McGovern TH, . . . Smiarowski K (2016) Climate challenges, vulnerabilities, and food security. Proc Natl Acad Sci 113(2): 298–303

Okpara UT, Stringer LC, Dougill AJ (2017) A novel climate – water conflict vulnerability index to capture double exposure: application in the Lake Chad Basin. Reg Environ Chang 17(2): 351–366

Omi M, Winant H (2015) Racial formation in the United States, 3rd edn. Routledge, New York

Ruppert T, Deady EL (2017) Climate change impacts on law and policy in Florida. In: Florida's climate: changes, variations, & impacts. Florida Climate Institute, Gainesville, pp 209–234

Sadri AM, Ukkusuri SV, Gladwin H (2017) The role of social networks and information sources on hurricane evacuation decision making. Natural Hazards Review 18(3):04017005

Scott M, Gallent N, Gkartzios M (2019) Planning rural futures. In: Scott M, Gallent N, Gkartzios M (eds) The Routledge companion to rural planning. Routledge, New York

Sultana F (2014) Gendering climate change: geographical insights. Prof Geogr 66(3):372–381

Sultana F (2017) Gender and water in a changing climate: challenges and opportunities. In: Water security across the gender divide, vol 93, 2nd edn. Springer, Cham, pp 17–33

Tall A (2019) Climate forecasting to serve communities in West Africa. Procedia Environ Sci 1:421–431

Umoh GS, Udo EJ, Solomon VA, Edet GE, Uwem CA, Okoro G, Bassey N, Akpan DO, Atairet E (2013) Adaptation to climate change. In: Agricultural ecosystems and gender dimensions. http://www.xlibrispublishing.co.UK

UN (2018) Scoping paper for expert group meeting on climate variation adaptation. African minister conference on the environment 12th meeting of the expert group TBC

United Nations Children's Fund (UNICEF) (2011) Situational analysis of children and women in Mongilia. UNICEF, Ukanbaatar

White I, Howe J (2012) Exploring the social impacts of flood risk and flooding in Scotland. Research findings no 32. Scottish Executive Social Research, Edinburgh

World Health Organization & World Bank (2011) World report on disability. Geneva, Switzerland: World Health Organization

11

# Impacts of Climate Change to Poultry Production in Africa: Adaptation Options for Broiler Chickens

M. O. Abioja and J. A. Abiona

## Contents


Introduction .
Climate Change and Its Evidences in Africa
Impacts of Climate Change to Poultry Production
  Direct Impacts
  Indirect Impacts
Broiler Production
  Heat Stress and Broiler Chickens
  Negative Effects of Heat Stress in Broiler Chickens
Adaptation Options to Climate Change for Broiler Chickens
  Housing System and Climate Amelioration
  Feed and Feeding Manipulations
  Water Supply
  Stocking Density
  Epigenetic Adaptation to Thermal Assaults ....
  Use of Honey and Other Natural Phytochemicals


M. O. Abioja (✉) · J. A. Abiona
Department of Animal Physiology, College of Animal Science and Livestock Production, Federal University of Agriculture, Abeokuta, Nigeria
e-mail: abiojamo@funaab.edu.ng; abionaja@funaab.edu.ng

Breeding for Thermotolerant Strains in Africa .
Others Adaptation Options
Conclusion
References

**Abstract**

Global climate change poses a great threat to poultry production. Greenhouse gases (GHGs) are released through both natural and anthropogenic sources into the atmosphere. Though poultry production contributes little to the release of GHGs, the subsector has been shown to be greatly affected by climate change and global warming. Poultry production as a major subsector of agriculture has provided the teeming population with a supply of needed animal protein in terms of meat and egg production all over the world. It is yet a major global employer of labor. Though it occupies a vantage position in meeting human needs, it is being threatened by climate change, especially in Africa where necessary structure to tackle the menace is nonexistent. Broiler chickens that are reared mainly for chicken meat cannot tolerate the high ambient temperature that prevails mostly in the tropical environment. Chickens are homeotherms that homeostatically regulate core body temperature within a narrow range. Elevated ambient temperature above thermal comfort zone, such as envisaged in climate change scenarios, will trigger series of neuroendocrine modulations that are detrimental to the welfare and productivity in broiler chickens. Such birds are said to be undergoing heat stress (HS). Negative effects of HS include reduced feed consumption, growth rate, feed digestion and efficiency, immunity, welfare, and survivability. Various adaptive measures that could be harnessed by broiler farmers, ranging from housing, feeding, watering, stocking, breeding for thermotolerant strains, thermal conditioning, use of phytochemicals, and much more, are reviewed upon in this chapter.

**Keywords**

Acclimation · Acclimatization · Broiler chickens · Environmental temperature · Heat stress · Thermal comfort zone · Thermoregulation

## Introduction

Climate change is defined by Ngaira (2007) as deviations in the patterns of climate over a long period of time while the Intergovernmental Panel on Climate Change (IPCC) recognized it as a change in the state of the climate that can be identified (using statistical tests) by changes in average and standard deviation, and that persists for a long period of time, typically more than one decade or longer (IPCC 2007). It poses a great threat to agricultural and socioeconomic development (Niang et al. 2014). The problem of climate change will add to the burden of heat stress (HS) usually experienced in the tropical regions of Africa because of the predicted increase in global temperature. Though Africa contributes the least of the global GHG emission, yet evidences had shown that most developing countries in Africa would be the most affected because of low infrastructural capacity to cope. This is also hinged on the fact of widespread poverty, prevailing slash-and-burn agriculture, erosion, and burning of firewood and farm residues. The problem of desert encroachment and deforestation is also a contributory factor to climate change. All facets of human life will be affected by climate change. Poultry production, a subsector of agriculture that had helped in supplying the needed human protein requirements from chicken, turkey, duck, guinea fowl, geese, and ostrich is not spared. Productivity and welfare of poultry species in Africa will be negatively impacted upon by climate change. Chicken being the most populous and important among poultry species will be most impacted upon. Global chicken population was over 22 billion in 2017. Broiler chickens, raised mainly for meat production with high feed-to-meat conversion efficiency, is the most sensitive to the effects of elevated temperature in the environment among other strains of chickens. Its productivity is grossly affected once the required environmental conditions, especially temperature, are compromised. These necessitate the development of a review of all available options for adaptation of broiler production to the prevailing and envisaged global warming.

## Climate Change and Its Evidences in Africa

Livestock subsector was viewed as a victim until it was implicated as a major contributor to global emission of greenhouse gases (GHGs) such as $CO_2$ and $CH_4$. FAO's report *Livestock's Long Shadow* opened up the minds of many stakeholders to this fact (FAO 2006) that the subsector contributes 18% of global GHGs. These gases cause warming of the globe by entrapping heat on the earth crust not allowing it to be reflected to the atmosphere. Resultant global warming and climatic variability have been reported to have both direct and indirect impacts on livestock production, including reduced growth and reproductive efficiency, low quality and quantity of feed materials, and increased prevalence of disease due to rise in temperature (Renaudeau et al. 2012). Africa is the most threatened all over the world with predicted effects of climate change (Ngaira 2007). Though, the continent is the least emitter of these GHGs of all continents, unlike other industrialized nations that are emitting these in tonnes daily.

Predicted temperature rise of 2–6 °C is expected over the land of Africa within the next 100 years and a rise of 1.5–3.0 °C by 2050. This is much more severe than experienced in other regions. The presence of climate change in Africa is evident in that the three warmest years over African land since 1950 had been identified as 2010, 2016, and 2017, all occurring within the last decade. Average deviation from the mean temperature of 1961–1990 was +1.41, +1.26, and + 1.20 °C, respectively. The warming rate per century recorded in Africa (+3.7 °C) is an alarming signal. ACMAD (2017) stated that temperature anomalies varied over the different subregions of Africa in 2017. Northern Africa experienced the average temperature anomaly was 1.1 °C warmer than the long-term mean. Similar to the experience in the North Africa, 2017 had temperature anomaly of 1.2 °C above average over the West African region. Central Africa had a temperature anomaly of 1.3 °C above average. Over Eastern Africa land, 1.2 °C was recorded above average while anomaly of 1.0 °C above average was reported over the Southern Africa region in 2017. However, temperatures in 2017 were mild over the island countries in the Indian Ocean. For instance, one of the coldest years in Madagascar is 2017, with temperature deviation of 0.3 °C below the mean. Climate change will pose considerable risks to the livelihood of the dwellers of most countries in Africa that mainly depend on agriculture (Hummel 2015).

Poultry farmers in Africa are aware of climate change because many of them have been taken cognizance of variability in rainfall amount and pattern, heat spells, and variability in other climatic elements. Indigenous technical knowledge and systems that had been explored by farmers in adapting to high environmental temperature abounds.

## Impacts of Climate Change to Poultry Production

In Africa, there are two main poultry production systems. Distinguishing factors between the two are associated with the scale of production, stock, management system, and productivity. The two systems are the commercial poultry and the rural poultry. The former is on large to medium scale, stocking improved/exotic breeds, reared intensively with adequate care for feeding, health, and welfare of the birds in modern facilities in large number. In terms of productivity, the commercial gives higher returns. On the other hand, the latter, *also known as* village or backyard poultry, is usually on small scale, stocking the locally adapted indigenous breeds of birds. The productivity is lower with the village poultry system. In history, the latter is far older than the former. Meanwhile, intermediate system between the two systems has evolved over time. Commercial poultry system includes breeding farms where grandparent (GP) and parent stocks are raised. Day-old chicks (broiler and layer lines) are produced by the breeding farms. Commercial broiler farms raise the chicks till table size of around 2.5 kg in 6–8 weeks. The effort to genetically improve commercial stocks of broiler chickens began in earnest in the USA in the 1940s (FAO 2020). And since then, it has been the stories of great leaps in improvement of growth rate and feed conversion efficiency. The successful breeding company started exporting chicken hybrids with improved strains into African countries in the 1950s. This coincided with the rapid development of commercial poultry systems in Africa.

All these systems are under the threats of the climate change but the most affected of them is the commercial poultry system. Impacts of climate change on poultry production (Fig. 1) could be categorized into two.

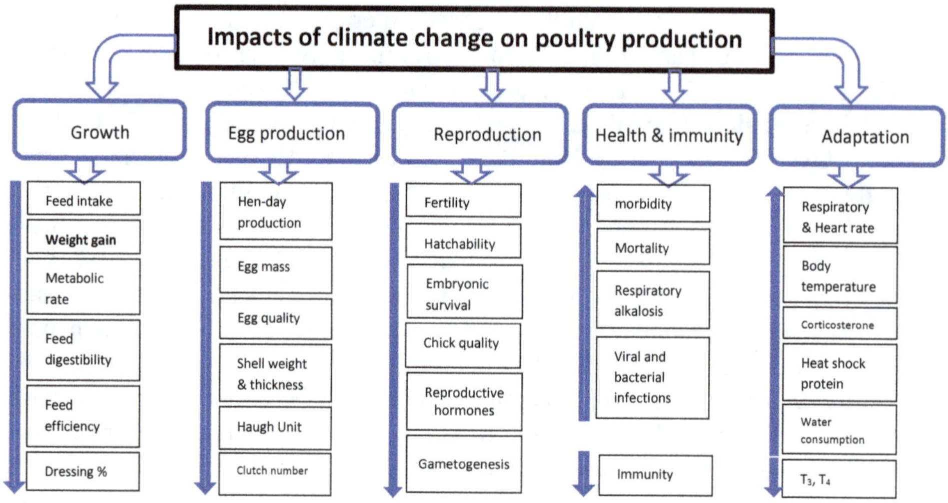


**Fig. 1** Diagram illustration the effects of climate change on poultry production

## Direct Impacts

Climate change has impacts on poultry production by way of imposing stress on the homeostasis in the birds. It may come in the form of extreme climatic situations: elevated temperature, flood or drought, and water scarcity (Tiruneh and Tegene 2018). Common among these is increased global temperature. Extreme weather conditions lead to production losses (reduced growth rate, lowered egg hen-day production, and increased morbidity and mortality) in poultry birds (Attia et al. 2011). Poultry birds can adapt to hot environment. However, the mechanisms of coping subject the birds to losses and diversion of nutrients meant for production to thermoregulation.

Growth performance of the birds is grossly affected by perturbations in weather conditions in the surroundings. The most important of these is the pen temperature. Changes in humidity work synergistically with high environmental temperature to impact negatively on poultry birds. In hot environment, feed consumption is reduced in order to limit metabolic heat production. Available nutrients in the system are utilized for heat loss mechanisms instead of profitable usage for muscle accretion. Digesta mobility in the gastrointestinal tract (GIT) is drastically reduced and efficiency of nutrient digestion and assimilation in the intestines is lowered. Histomorphometric studies show that the epithelial cells of the intestines are affected by HS (Santos et al. 2015). Vasodilation of blood vessels at the periphery reduces blood flow across GIT of the birds thereby lowering the efficiency of digestion and nutrient uptake. All these factors cumulatively contributed to the lowered final liveweight and dressing percentage usually obtained in heat-stressed chickens (Syafwan et al. 2011).

Productivity in laying chickens is negatively affected by elevation in temperature by reducing the number and size of eggs produced. Internal and external quality traits of the eggs are compromised by changes in weather conditions around the birds. High ambient temperature lowers egg fertility, hatchability, and chick quality in breeder stocks (Ayo et al. 2011). Impact of climate change will be felt on the synthesis and the release of reproductive hormones (FSH, LH, progesterone, estrogen, and testosterone) with resultant reduction in efficiency of gametogenesis in male (spermatogenesis; Al-Saffar and Rose 2002; Karaca et al. 2002) and female (oogenesis) birds (Rozenboim et al. 2007). Heat-stressed cocks produce semen with low sperm concentration and quality. Immunity of the birds is lowered under HS conditions (Calefi et al. 2017).

## Indirect Impacts

Indirectly, climate change will impact on feed ingredient availability and quality for the poultry birds; availability of adequate good quality water; and pest and diseases infestation in Africa. Climate has effects on the yield and quality of produce of crop husbandry from where most of the feed ingredients are sourced. Feed ingredients such as maize, groundnut cake, cotton-seed cake, wheat offal, and rice bran are gotten from agronomic activities. Semiarid regions experiencing low rainfall amount and irregular pattern cannot boast of bountiful harvest. This may trigger skyrocketing of the feed price for poultry species and exacerbate the animal-human competition for feed materials except irrigation systems are employed. Rain-fed cropping in the forest vegetation will also be affected by high temperature and insufficient rainfall.

Availability of adequate good quality water is essential to productivity in poultry production. Climate change has reared up its head in reduction in water in the ponds, streams, rivers, and the seas immediately after rainy season all around Africa because of increased rate of evaporation (Alemayehu and Woldeamlak 2017). Water shortage adversely affects body weight and lymphoid organs (Mustafa et al. 2010). Changes in the existing pattern of pest and disease infestations are envisaged under different climatic change scenarios. This will affect productivity of poultry in terms of morbidity, mortality, and the cost of vaccinations and medications.

## Broiler Production

Broiler chickens are reared mainly for meat production. The production cycle lasts for 5 to 8 weeks, with high rate of turn-over of feed to meat from day-old chicks to market age. Chicks are obtained from specialized lines called breeder chickens. At market age, the liveweight of 40–55 g at day-old would have multiplied by 50 times to give about 2.5 kg chickens. Both male and female sexes of broiler chickens are reared for meat, though male chickens grow faster than the female counterparts. Most of the strains of broiler chickens reared in Africa are developed in the temperate regions of the world such as the USA and the Netherlands but are imported into Africa. The birds are incapable of expressing the inherent growth potentials because of HS and the productivity is often below what is obtainable in their counterparts in the cooler climes. Broiler chickens are sensitive to elevation in ambient temperature during growing phase. Production and demand for chicken meat has increased tremendously worldwide over the last few decades. This results from different selection and cross-breeding techniques that are employed to improve growth rate of broiler chickens, reaching market weight within a short span of time than what is obtained in the past. To continue enjoying the gains of several years of broiler development, efforts toward ensuring adaptation of broiler chickens to prevailing high environmental temperature expected in climate change are worthwhile.

## Heat Stress and Broiler Chickens

All farm animals in the tropical region suffer from HS at one time or the other because of constant elevated ambient temperature (Altan et al. 2000). People and livestock in the temperate regions are exposed to HS induced by elevated environmental temperature during summer. HS is an adaptive response that occurs in a bird when the rate of thermolysis is below thermogenesis and the ability to lose body heat exceeded by the heat load acquired through exposure to high ambient temperature (Al-Saffar and Rose 2002).

Broiler chickens are homeothermic animals capable of maintaining the body temperature within a narrow range irrespective of the environmental temperature. They possess an internal homeostatic mechanism that regulates internal body temperature. The internal body temperature of adult chicken is normally between 41.2 °C and 42.2 °C (Mitchell et al. 2005). Newly hatched birds have a body temperature approximately 2–3 °C below that of the adult birds. Additional source of warmth is needed by the chicks during the first 21 days post-hatch to maintain body temperature for normal growth and development. However, as the birds increase in age and size, their requirement for supplemental heat declines. This results from the development of insulating feathers, higher metabolic heat production, and maturation of the thermoregulatory system of the birds. In general, the thermal comfort or thermoneural zone (TNZ) for broilers declines from about 32 °C at hatching to around 24 °C at 3 to 4 weeks of age and to about 21.1 °C thereafter. The birds at this period are capable of heat regulation to maintain core body temperature. As temperature increases beyond 21.1 °C around adult chickens, mechanisms for heat loss are triggered.

The fast-growing broiler lines are susceptible to high temperature mostly during growing-finishing phase. The high susceptibility to HS results from resultant inferior development of their cardiovascular and respiratory systems (Yahav 2000). As well, chickens do not possess sweat glands (Mitchell et al. 2005). In addition, their rapid growth rate is supported by high feed intake, thus, as they grow, metabolic heat production increases but their heat dissipation capacity does not. Broiler chickens gain heat from metabolism, physical activity, and environment. This heat load must be dissipated to maintain constant body temperature. Metabolic heat production in broilers is particularly high because of their high feed intake compared to other strains of chickens. Their growth rate is supported by feed consumption which leads to generation of heat in the body system. Increased heat production does not have

**Fig. 2** Diagram showing broiler chicken exhibiting panting behaviour

any adverse consequence on the birds in TNZ and cold environments. But as ambient temperature overshoots the upper limit of the TNZ, the ability of the bird to dissipate heat is compromised making excessive heat production built-up. The condition leads to hyperthermia (increased core body temperature) which is potentially life threatening. Before this stage, the body temperature is regulated mainly by loss of heat to outside environment through conduction, convection, and radiation, *also known as* non-evaporative heat loss. Effectiveness of these three means of heat loss is limited to lower temperatures (Borges et al. 2003).

As the temperature peaks, chickens exhibit panting, an evaporative heat loss behavior (Fig. 2; Video 1). Panting involves evaporative cooling of the bird by losing heat from the respiratory tract. They also display wing-raising to expose the poorly feathered sides of the body and under-wing area. Sometimes, they dig into the litter to sit on a cooler floor in the pen during hot periods. Panting occurs when the deep body temperature of poultry reaches 42 °C. Respiratory rate may increase from 25 to 150 breaths per minute over a 20-min period in response to an increase in ambient temperature from TNZ. In a healthy chicken, hyperventilation through panting will remove approximately 0.54 Kcal/g water lost in the lungs. Chickens increase saliva secretion during this period because of the need to the surface of the respiratory tract wet during panting.

## Negative Effects of Heat Stress in Broiler Chickens

The adverse effects of HS can be seen in decreased feed consumption, increased water intake, rise in body temperature, respiratory rate, heart rate, electrolyte imbalance, changes in hematological parameters, hormone levels and enzymatic activities, and perturbations in blood *pH* (respiratory alkalosis; Lara and Rostagno 2013). All these negatively affect productivity of the animal. An increase in body temperature above the regulated range, because of exposure to hot environmental conditions, may lead to a cascade of irreversible thermoregulatory events that may be lethal for the birds. Generally, broiler chickens try to lower its heat production by reducing its feed consumption during heat episodes. The reduction in feed consumption can be as high as 1.5% per 1 °C increase in temperature above TNZ. Reduction in feed consumption consequentially leads to insufficiency of essential nutrients. Coupled with this is the lowered digestibility of feed in the gastrointestinal tract of the birds. Growth rate is therefore reduced in broiler birds when environmental temperature rises because even the little energy obtained from the small feed consumed is expended in panting. The result is that birds had lower final body weight. Chronic HS increases the time to reach market weight and impairs feed digestibility. More feed is required to lay down unit weight of chicken than in TNZ conditions. Exposure to hot weather leads to high mortality in broilers and loss of immunity. Efforts to increase feed consumption during HS by force-feeding have been shown to decrease survival. Mortality in broilers up to 10% of the total production has been reported during HS. Egg production, fertility, and hatchability are said to be grossly affected in both grandparent and parent (broiler breeder) stocks. Time expected for feeding is expended in panting and beside water troughs. Birds consume more water during HS exposure. Excessive loss of $CO_2$ in exhaled air during panting results in high blood *pH*, a syndrome called *respiratory alkalosis*.

## Adaptation Options to Climate Change for Broiler Chickens

### Housing System and Climate Amelioration

Ideal broiler chicken environment must meet the requirement of the birds for adequate pen temperature, relative humidity, and air circulation. Broiler houses are closed system in most industrialized countries where the temperature and humidity are fully controlled, unlike in developing countries where open-sided housing units are common. Foggers, tunnel ventilators, fans, and misters are used to cool the environment in closed systems. However, the cost of installing and maintaining fans, foggers, ventilators, misting pumps, and so on in poultry houses is often beyond what most farmers can afford.

Attaining the optimal range of climatic variables in the open-sided poultry houses common in the tropics is difficult and almost impossible (Abioja 2010). However, proper ventilation is essential in minimizing the effect of heat in African poultry housing units. As the higher environmental temperature is expected all over Africa, the following should be considered for broiler housing:

- Orientation of the housing units should allow for cross–ventilation.
- Low-walled structures completed with wire mesh to ensure cross-ventilation (Fig. 3). High walled structures trap heat inside and will not be applicable.
- Mud wall houses provide cooler interior than brick wall housing unit.
- Roofing style should be such that allow for proper ventilation.
- Roofing materials must be considered. Asbestos roof is preferred to corrugated iron sheet. Local materials that enable heat transfer such as palm fronds, long grasses such used for thatched roof could be used.
- Tree planting around broiler housing units will provide coolness (Fig. 4).
- The use of mobile pens with wire mesh floor placed under shade of trees.

**Fig. 3** Low-walled broiler housing units

**Fig. 4** Tree shade beside broiler housing unit

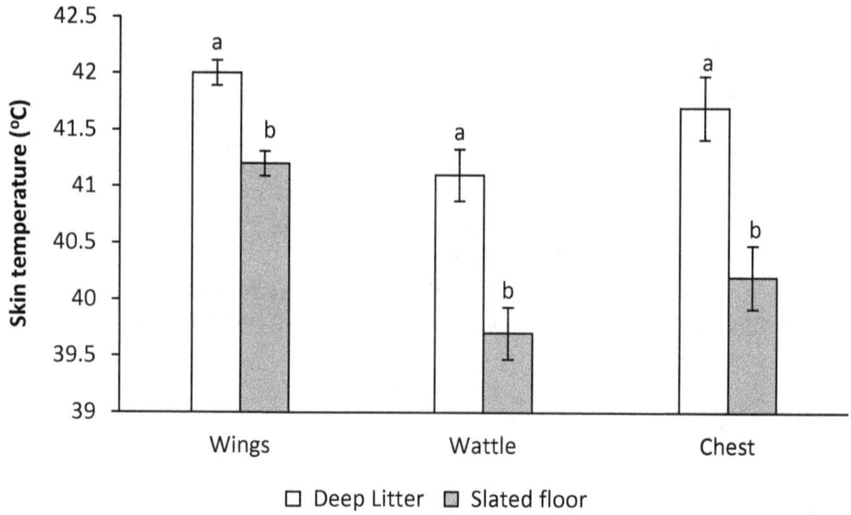


**Fig. 5** Effect of floor type on skin temperature of broiler chickens during hot-dry season

**Table 1** Effect of floor type on rectal temperature, respiratory rate, heart rate, and skin temperature in broiler chickens during hot season

| | Floor | | |
|---|---|---|---|
| Parameter | Deep-litter | Wire mesh | *p* |
| Rectal temperature, °C | $42.7 \pm 0.07^{a}$ | $42.0 \pm 0.07^{b}$ | 0.000 |
| Respiratory rate, breaths/min | $138.8 \pm 2.28^{a}$ | $125.3 \pm 2.28^{b}$ | 0.000 |
| Heart rate, beats/min | $198.4 \pm 2.96^{a}$ | $188.6 \pm 2.96^{b}$ | 0.023 |

$^{a,\,b}$Means within the same row with different superscripts differ significantly ($P < 0.05$)

Broiler chickens are commonly raised on deep-litter floor in most areas of the world, but the use of cage for broiler is becoming common (Shields and Greger 2013). Type of floor has been reported to influence liveweight, feed intake, protein efficiency, and feed conversion ratio in broiler birds reared during winter and summer (Simsek et al. 2014). Wire mesh floor decreased skin temperature, cloacal temperature, respiratory rate, and heart rate of broiler chickens compared to deep-litter floor during hot season (Fig. 5, Table 1; Abioja 2020 –*Unpublished data*).

## Feed and Feeding Manipulations

Special attention should be given to the nutrition of broiler birds under high temperatures expected under climate change. The major problem of broiler chickens under HS conditions is the fact of reduction in feed consumption. Reduced dietary intake limits available metabolizable energy to sustain normal output. The following suggestions may be of help in feeding broilers during heat spells:

- More energy should be delivered in the form of fat instead of starch (because heat increment from metabolizing fat is lower than that from carbohydrate however rancidity of fat should be given a consideration).
- Balancing of amino acids in the diet especially methionine and lysine is necessary to augment for deficiencies resulting from low protein intake.
- Diets with low protein level should be recommended in order to reduce heat production in broiler chickens under HS because protein has the highest heat increment of all nutrients.
- Anti-oxidant vitamins (A, C, and E) supplementation has also been credited to improve performance of broilers during HS.
- Wet feeding (Awojobi et al. 2009; Dei and Bumbie 2011) was found to help in improving the final liveweight and weight gain of the birds in a hot climate. Syafwan et al. (2011) stated in a review on HS and feeding strategies in meat-type chickens that wet feeding may be profitable under HS conditions. Feeding wet diets might facilitate an increased water intake. This ensures availability of adequate water for evaporation during panting, hence helping cooling of the bird.

Other areas to be considered for adaptation of broiler chickens to climate change include feeding time (Farghly et al. 2019). Feeding time should be organized in a way that the peak metabolic heat production will not coincide with the highest point in environmental temperature.

## Water Supply

Water is one of the limiting factors in broiler production. It is an essential nutrient that must be taken into consideration if expected productivity is to be achieved. Factors that are of importance in drinking water for chickens include quantity, quality, temperature, salinity, and microbial load. Water supply to the broiler houses must be clean and constant. Broiler chickens consume more water during hot period than in cold climate (Manning et al. 2007; Bruno et al. 2011). Its restriction was reported to raise heterophil-lymphocyte ratio, a major indicator of stress in broiler chickens. Availability of drinkable water during hot spells must be ensured. Water restriction caused reduced weight of lymphoid organs in broiler chickens (Mustafa et al. 2010), which may negatively impact the innate immunity of the birds. In South-Eastern Nigeria located in the rain forest zone, water restriction up to 1.2 liters per day was found to have deleterious effects of the hematological parameters and caused increased activity of liver enzyme which is an indicator of cellular necrosis in broiler chickens (Iheukwumere and Herbert 2003). Water restriction had negative effects on the renal functioning, digestion, immunity, and other vital body systems. Birds subjected to water restriction had an altered behavior, becoming aggressive and irritated. Therefore, water restriction irrespective of the magnitude must not be allowed in broiler production.

Water temperature is another important factor that must be put into right perspective. Warm water is nauseating and not readily acceptable to chickens.

**Table 2** Rectal temperature and respiratory rate in broilers offered water of two different temperatures

| Parameter | Water temperature | | Standard error of means (sem) |
|---|---|---|---|
| | Ordinary | Low | |
| Water temperature (°C) | 30.0 | 16.0 | 0.09 |
| RT at 08.00 h (°C) | 41.7[a] | 41.5[b] | 0.03 |
| RT at 16.00 h (°C) | 42.5[a] | 42.1[b] | 0.04 |
| RR at 08.00 h (breaths/min) | 54.0[a] | 46.7[b] | 1.13 |
| RR at 16.00 h (breaths/min) | 85.2[a] | 62.6[b] | 2.11 |

Abioja MO (Unpublished data)
[a, b]Means with differing superscripts within a row were significantly different ($P < 0.05$)

Chilled water is preferred to water at ambient temperature. Offering water at low temperature may reduce heat load and respiratory rate in broiler chickens (Table 2). This has been proved to help combat HS in broilers (Abioja et al. 2011, 2013). Practical activity in cooling water may involve having water reservoirs placed under shade to disallow overheating by direct sunlight during hot periods. Many poultry managers in all over Africa have been supplying their birds with water cooled with ice cubes during the dry season. The enhanced respiratory rate during HS is critical for body temperature maintenance because of its resultant evaporative cooling.

Specific consequences of the acid-base perturbations have indeed been identified. Electrolytes lost can be replenished with sodium (Na), potassium (K), calcium (Ca), magnesium (Mg), chlorine (Cl), bicarbonate ($HCO_3$), and sulfate ($SO_4$) salts in water (Ahmad and Sarwar 2006). The addition of the electrolytes to the drinking water not only alters the bird's osmotic balance by replenishing those depleted during HS but also stimulates water consumption and influences water balance. Some African native chickens had been found to have tolerance for water shortage. Chikumba and Chimoyo (2014) reported that Southern African naked neck chickens was found to performed better than Ovambo chickens under conditions of water restriction and would be ideal to raise for meat and egg production in locations where water shortages are a major challenge.

## Stocking Density

Each bird in the flock emits radiant heat to the environment. Low stocking density reduces the number of birds producing heat and the amount of heat that must be removed from the house to maintain temperature. Recommended stocking density appropriate for different housing type should be strictly adhered to. Assuming each bird weigh 2 kg at slaughter, for controlled environment, the maximum stocking density at slaughter should be 30 kg/m$^2$; controlled environment during hot period, 24–26 kg/m$^2$; open-sided house, 20–22 kg/m$^2$; and open-sided house during hot period, 16–18 kg/m$^2$. Abudabos et al. (2013) reported that increasing the stocking density rate from 28 to 40 kg of body weight/m$^2$ had evident negative effects on the growth performance of broiler chicken. Higher stocking density does jeopardize the well-being of the chickens.

## Epigenetic Adaptation to Thermal Assaults

Chickens are precocial birds having the development of body functions starting early during the embryonic phase (Nichelmann and Tzschentke 2002). Thermoregulatory mechanisms start during prenatal stage but not fully matured as at pre- and peri-hatch periods (Tzschentke and Basta 2002; Tzschentke et al. 2004). At this point, exposure to a level of stress can trigger an imprinting in chickens. Acute changes in the environmental conditions especially thermal environments induce as a rule, first uncoordinated and immediately nonadaptive reactions. Later, the uncoordinated (immediately) nonadaptive reactions change into coordinated (adaptive) reactions (Tzschentke 2007). Stress is usually avoided because of its perceived repercussions on living tissues. However, evidences are emanating from various research works that exposure of broiler chicks to mild stress at early age help in acquiring thermotolerance at latter age. This was first reported in rats by Levine (1962). The author discovered that mild stress at early age affects the adrenocortical functioning during adulthood.

It was later confirmed in broiler chickens (Yahav and Plavnik 1999; Zulkifli et al. 2000; Liew et al. 2003). From this assertion, there are two strategies that have been explored: thermal conditioning (TC) and early feed restriction (EFR). TC involves exposing chicks at early age post-hatch to elevated ambient temperature (Arjona et al. 1990; Yahav and Hurwitz 1996; Yahav and McMurtry 2001) within the first 6 days while EFR subjected the chicks to varying degrees of hunger during the same period (Zulkifli et al. 1994). TC has been demonstrated to be effective as a tool for thermotolerance acquisition in poultry but its application is not practicable in the traditional open-sided poultry houses common in most tropical and developing countries. The cost of heating poultry house to a desired temperature is high. Besides, the technicality associated with thermal conditioning such as length and degree of heat exposure may not be easily mastered by the local farmers. This leaves the only option for thermotolerance acquisition in the developing and underdeveloped world to application of early feed restriction.

Sublethal stress imposed by feed withdrawal in the early age post-hatch on broiler chicks has been noted to confer inherent capability in the birds to cope with stress in latter age (Liew et al. 2003). Abioja et al. (2014) have found that applying EFR to broiler chicks of d5 post-hatch for 24 h may help reduce hyperthermia that is common during heat spell in pen at market age. It is not known whether the response will be the same during transportation as in the acute heat spell. Improved thermotolerance in broiler chickens at market age will help against losses that do occur during transportation. Adopting feed restriction during early period of life in broiler chickens as a means of improving thermotolerance in broiler chickens is easier for farmers than adjusting pen temperature. Induced thermotolerance is generally referred to as the state at which an organism is transiently more resistant to killing by heat due to a short pretreatment at moderately elevated ambient temperatures. This usually takes place before the brain is fully matured for thermoregulation. During ontogeny in chicken, embryo undergoes a transition from poiklothermy to homeothermy, which is completed, at the early post-hatching. Full-blown homeothermy starts at approximately 10 days post-hatch. Early age thermal conditioning by exposing young chicks to 40 °C for 24 h reduces body temperature and improves long-term broiler resistance to HS without negative effects on growth and feed conversion ratio.

Zulkifli et al. (2000) reported that acute HS resulted in increase in heterophil/lymphocyte ratio for all feed-restricted groups and the ad libitum group. But broilers restricted from 60% of daily feed requirement on day 4, 5, and 6 had the least heterophil/lymphocyte ratio. The authors concluded it appeared 60% feed restriction is beneficial in improving growth and survivability of female broiler chickens exposed to HS later in life. The authors restricted the feed by percentage on three consecutive days (day 4, 5, and 6). Exact feed intake of the birds may not be ascertained: feed intake depends on strain, genotype, environmental condition in the pen, body weight, etc. Farmer may have problem calculating the feed intake. Besides, percentage feed restriction may not give the exact day the EFR is most effective. Therefore it is important to determine exact day of life the chicks should be subjected to early feed restriction by farmer which ensures best acquisition of thermotolerance.

## Use of Honey and Other Natural Phytochemicals

The age-long conventional use of antistress antioxidant formula containing a mixture of vitamins and minerals to help broiler chickens cope during heat spell in rearing has proved to be effective. However, the use of phytochemicals that possess or are suspected to possess efficacious antioxidant properties in the management of HS in poultry production is becoming more popular (Abioja et al. 2012). Efficacy of honey and some other naturally occurring materials from plants such as coconut water, olive leaf extract, and propolis during stressful conditions in chickens had been reported in literatures. Honey has recently been reputed as a possible natural source of antioxidant in broiler chickens (Abioja et al. 2012; Osakwe and Igwe 2015; Oke et al. 2016) with some positive results because of its phenolic and flavonoid compounds content and various phytochemicals that possess antioxidant properties. It has been used during acute heat and transportation stress conditions. Honey is effective because some of its components are substances known to have physiological actions in the body systems of animals. Diluted drinking water of concentration 20 ml honey dissolved in a liter drinking water had positive effect on respiratory and heart rates, calcium metabolism, bone formation, and some internal organs in heat stressed broiler chickens. Honey in drinking water could be of help in improving the welfare of broiler chickens during stress episode by increasing the packed cell volume, red blood cell count, and hemoglobin concentration according to Abioja et al. (2019). It helps in protecting gastrointestinal tract, liver, kidney, pancreas, eye, plasma, red blood cells, and reproductive organs against oxidative stress in rats (Zaid et al. 2011; Erejuwa et al. 2012).

The presence of phenolic compounds and flavonoids in honey had been primarily implicated as the reason for its antioxidant activity. Phenolic compounds are good in free radical scavenging, hydrogen donation, singlet oxygen quenching, and metal ion chelation. Phenolics and flavonoids do help in correction of the redox perturbations in the body by counteracting the damage caused by oxidants such as oxygen, hydroxide, superoxide, and/or lipid peroxyl radicals. Exogenous nonenzymatic antioxidants in honey synergistically support the endogenous antioxidant enzymes, such as superoxide dismutase (SOD), catalase (CAT), and glutathione peroxidase (GPX), to eradicate reactive oxygen species. Honey has various antioxidant constituents that act together for the same goal but at different cellular and subcellular levels (Erejuwa et al. 2012; Oryan et al. 2016; Kamaruzzaman et al. 2019).

It contains no fat and cholesterol contents. However, the use of honey must be with caution. It was reported that there were infiltrations of fat cells in the liver tissue of albino rats fed with honey over a long period. This may result in nonalcoholic fatty liver disease or in other unpleasant conditions harmful to health if not well managed. It has been discovered that long-term administration of honey caused deformation and atrophy of specific reproductive apparatus in laying chickens. Other authors had earlier stated that chronic consumption of unprocessed Nigerian honey resulted in decreased bile flow, increased bile cholesterol, and decreased plasma cholesterol in albino rats. The fear that more cholesterol molecules especially much dreaded low-density lipoproteins may be deposited in the eggs of chickens offered honey is pervading the air. Besides, vitamin A content is important in human nutrition. Improvement in vitamin A content of eggs will go a long way in correcting its deficiency and contribute to the welfare of consumers. Millions of preschool-aged children in developing countries are suffering from vitamin A deficiency, which may lead to blindness (Underwood 1998). The deficiency of vitamin A is a public health issue that needs attention. Fortification of eggs for the children will be a welcome idea, should honey help in achieving this. By-products of honey such as honey slum gum (Babarinde et al. 2011), propolis (Chen et al. 2009), and bee pollen (Wang et al. 2005; Haščík et al. 2012) had been used in broiler production and found to be effective.

Honey is good but must be used with caution, especially in its application in poultry production. In literature, it has been reported that there were infiltrations of fat cells in the liver tissue of rats which was dosage-dependent (Avwioro et al. 2012), decreased bile flow, increased bile cholesterol, and decreased plasma cholesterol (Alagwu et al. 2009) in Albino rats and hypertrophy of follicle, giving it a deformed shape instead of the usual round shape in laying chickens (Abioja and Adekunle 2018). The authors concluded that chronic consumption of honey may increase the risk of hepatic damage.

Naturally occurring solution found inside the fleshy eatable part of coconut called coconut water could also be used as antistress in broiler chickens during heat spells. It contains a lot of bioactive metabolites which have not been explored (Reddy and

Lakshmi 2014). Preliminary studies (Abioja et al. 2015) with diluted coconut water in drinking water as a viable candidate in lowering body temperature in heat and transportation stressed broiler chickens gave a promising results. Saat et al. (2002) stated that coconut water has been used in humans after physical exercise or long journey as an ideal rehydrating and refreshing drink. Its sugar content and mineral composition (electrolytes) have been identified as reason for the usefulness. Coconut water stimulates higher water consumption, which ensures availability of enough water in the body to facilitate evaporative cooling (panting) under hot environmental conditions. Chickens ingest more water under HS conditions. Prades et al. (2012a, b) reviewed the uses, composition, properties, preservation, and processing of coconut water. Major constraint in the adoption of coconut water as antistress in broiler production is in quantity that would be required for a large-scale commercial broiler industry. More research works are necessary to ascertain the appropriate concentration of diluted coconut water for the chickens. However, it can be used easily in small scale, backyard, and family poultry production. Other materials that could be used in ameliorating the negative effects of HS are raffia palm wine, olive leaf extract, and orange peel extract.

## Breeding for Thermotolerant Strains in Africa

Broiler chickens reared in the tropics were developed in the cooler regions of the world and introduced into the tropics. They often respond differently under the new environment, mostly growing at lower rate than in their home environment. Thus optimal poultry production in the hot regions therefore requires an adequate and appropriate management system that can reduce the effects of HS to the minimum. Olori (2008) in a review stated that to develop broiler chickens for Africa, an understanding of the factors affecting poultry production such as the climate, production systems, and available feed resources in Africa is required. One of the major constraints in broiler breeding is the fact that productivity traits are negatively correlated with thermotolerance. The available major genes (naked-neck, frizzle-feather, and dwarfism) in locally adapted breeds of chickens all over Africa must be explored. FUNAAB-alpha broiler lines were developed by Professor Olufunmilayo Adebambo and her team from several crosses and selection activities among flocks of Nigerian local chickens. The newly developed chicken lines had just been registered and tested in different agro-ecological zones of the Nigeria and parts of West African coast.

A recent report on the on-station performance evaluation of improved tropically adapted chicken breeds for smallholder poultry production systems in Nigeria was given by Bamidele et al. (2019). The authors in the study identified FUNAAB Alpha and Noiler as being more suitable for dual-purpose functions (egg and meat), while Sasso and Kuroiler (meat) and Shika-Brown (egg) were observed to be better suited for single purpose functions. Chickens bred taken into consideration the climate of African land should be given preference as the pangs of climate change is envisaged. Breeds like FUNAAB-alpha, Kuroiler, Noiler, Sasso, and Shika-brown chickens are now becoming popular in the hot regions of Africa.

## Others Adaptation Options

Organization of enlightenment programs, workshops, and trainings for the broiler producers and stakeholders on adaptation of broiler chickens to climate change from time to time by governmental agencies is necessary in Africa. Awareness and advocacy about climate change should be extended to the young people and other categories of people. Sustainable Development Goals should involve younger generations of African populace in schools where they can be furnished with adequate and up-to-date information about climate change. Indigenous technical know-hows of the African peoples, in forms of herbs, extracts, and other phytochemicals should be explored in keeping broiler chickens adapted to impending global warming.

## Conclusion

As the evidences of climate change are becoming more pronounced over Africa, various adaptation options available for broiler production in literatures include building housing systems that ensure climate amelioration for the birds, manipulations of feed and feeding systems, supplying adequate water of good quality, ensuring adequate stocking density, use of natural phytochemicals that ensure balanced oxidative status in the body system, and breeding for thermotolerant strains in Africa. Enlightenment programs should be organized for poultry farmers by governmental and nongovernmental agencies on the climate change mitigation, adaptation, and resilience strategies in order to improve livelihood of the people in a sustainable manner.

## References

Abioja MO (2010) Temperature-humidity effects on egg fertility and evaluation of vitamin C and cold water on broiler growth in hot season. A Ph.D. thesis submitted to Department of Animal Physiology, University of Agriculture, Abeokuta, 157pp

Abioja MO, Adekunle MO (2018) Climate change: depression in Egg production in chickens during the hot season with long term honey administration. In: Leal Filho W (ed) Handbook of climate change resilience. Springer International Publishing AG, https://doi.org/10.1007/978-3-319-71025-9_15-1

Abioja MO, Abiona JA, Sodipe OG, Adeleye OO, Akibo TE, Daramola JO, Oke OE, Williams TJ, Oguntuase OM, Oladapo OI, Okanlawon OR, Mudasiru AO, Siyanbola AY, Ogunduyi BO, Soyiga BM, James IJ, Osinowo OA (2015) Effect of coconut water on rectal temperature and blood glucose in broilers subjected to acute heat and transportation stress. Nig J Anim Sci 17 (2):141–147

Abioja MO, Akinjute OF, Balogun SI, Oguntimehin MS, Oluwasola TO (2019) Effect of honey on haematology, plasma biochemistry and liver enzymes in broiler chickens administered dietary corticosterone. Stress Physiol Biochem 15(3):5–13

Abioja MO, Ogundimu KB, Akibo TE, Odutola KE, Ajiboye OO, Abiona JA, Williams TJ, Oke OE, Osinowo OA (2012) Growth, mineral deposition and physiological responses of broiler chickens offered honey in drinking water during hot-dry season. Int J Zool 4:1–6

Abioja MO, Osinowo OA, Smith OF, Eruvbetine D (2013) Physiological and haematological responses of broiler chickens offered cold water and vitamin C during hot-dry season. Nig J Anim Prod 40(1):24–36

Abioja MO, Osinowo OA, Smith OF, Eruvbetine D, Abiona JA (2011) Evaluation of cold water and vitamin C on broiler growth during hot-dry season in South-Western Nigeria. Archiv de Zoot 60 (232):1095–1103

Abioja MO, Sodipe OG, Abiona JA, Oladipo KA, Kasali OD, Akerele Z, Ola A, Oke H, Ogundele O, Osinowo OA (2014) Thermotolerance acquisition in broiler chickens through early feed restriction: response to acute heat stress. Pertanika J Trop Agric Sci 37(3): 311–319

Abudabos AM, Samara EM, Hussein EOS, Al-Ghadi MQ, Al-Atiyat RM (2013) Impacts of stocking density on the performance and welfare of broiler chickens. Italian J Anim Sci 12:683–692

ACMAD (2017) The state of climate in Africa. African Center for Meteorological Application for Development. https://www.acmad.net/new/sites/default/files/The-State-of-Climate-in-Africa-2017-Report-March-2018-ACMAD.pdf. Accessed 11 Mar 2020

Ahmad T, Sarwar M (2006) Dietary electrolyte balance: implications in heat stressed broilers. Worlds Poult Sci J 62:638–653

Alagwu E, Nneli RO, Okwari OO, Osim EE (2009) Bile secretion in albino rat following chronic honey intake. Nig J Physiol Sci 24:203–206

Alemayehu A, Woldeamlak B (2017) Smallholder farmers' coping and adaptation strategies to climate change and variability in the central highlands of Ethiopia. Local Environ 22(7):825–839. https://doi.org/10.1080/13549839.2017.1290058

Al-Saffar AA, Rose SP (2002) Ambient temperature and the egg laying characteristics of laying fowl. Worlds Poult Sci J 58(3):317–332

Altan O, Altan A, Oguz I, Pabuccuoglu A, Konyalioglu S (2000) Effects of heat stress on growth, some blood variables and lipid oxidation in broilers exposed to high temperature at an early age. Br Poult Sci 41(4):489–493

Arjona AA, Denbow DM, Weaver WD Jr (1990) Neonatally-induced thermotolerance: physiological responses. Comp Biochem Physiol 95A:393–399

Attia YA, Hassan RA, Tag El-Din AE, Abou-Shehema BM (2011) Effect of ascorbic acid or increasing metabolizable energy level with or without supplementation of some essential amino acids on productive and physiological traits of slow-growing chicks exposed to chronic heat stress. J Anim Physiol Anim Nutr 95:744–755

Avwioro GO, Onyije FM, Atoni AD, Nduku A (2012) Non-alcoholic fatty liver disease following administration of unprocessed Nigerian honey. Adv Biol Res 6:141–145

Awojobi HA, Oluwole BO, Adekunmisi AA, Buraimon RA (2009) Performance of finisher broilers fed wet mash with or without drinking water during wet season in the tropics. Int J Poult Sci 8 (6):592–594

Ayo JO, Obidi JA, Rekwot PI (2011) Effects of heat stress on the well-being, fertility and hatchability of chickens in the northern Guinea savannah zone of Nigeria: a review. ISRN Vet Sci 1–10. https://doi.org/10.5402/2011/838606

Babarinde SA, Oladunjoye IO, Ojebiyi OO, Oyedeji SA (2011) Inclusion of honey bee slum gum in broiler chicken feed. Int J Agric Biol 13(5):781–785

Bamidele O, Sonaiya EB, Adebambo OA, Dessie T (2019) On-station performance evaluation of improved tropically adapted chicken breeds for smallholder poultry production systems in Nigeria. Trop Anim Health Prod. https://doi.org/10.1007/s11250-019-02158-9

Borges SA, Fischer da Silva AV, Ariki J, Hooge DM, Cummings KR (2003) Dietary electrolyte balance for broiler chickens under moderately high ambient temperatures and relative humidities. Poult Sci 82:301–308

Bruno LDG, Maiorka A, Macari M, Furlan RL, Givisiez PEN (2011) Water intake behaviour of broiler chickens exposed to heat stress and drinking from bell or and nipple drinkers. Brazilian J Poult Sci 13(2):147–152

Calefi AS, Quinteiro-Filho WM, Ferreira AJP, Palermo-Neto J (2017) Neuroimmunomodulation and heat stress in poultry. Worlds Poult Sci J 73:493–504

Chen YJ, Huang AC, Chang HH, Liao HF, Jiang CM, Lai LY, Chan JT, Chen YY, Chiang J (2009) Caffeic acid phenethyl ester, an antioxidant from propolis, protects peripheral blood mononuclear cells of competitive cyclists against hyperthermal stress. J Food Sci 74(6):H162–H167

Chikumba N, Chimonyo M (2014) Effects of water restriction on the growth performance, carcass characteristics and organ weights of naked neck and Ovambo chickens of southern Africa. Asian-Austral J Anim Sci 27(7):974–980

Dei H, Bumbie GZ (2011) Effect of wet feeding on growth performance of broiler chickens in a hot climate. Br Poult Sci 52(1):82–85

Erejuwa OO, Sulaiman SA, Ab Wahab MS (2012) Honey: a novel antioxidant. Molecules 17(4):4400–4423

FAO (2006) Food and agriculture organization of United Nations. Livestock's long shadow: environmental issues and options. http://www.fao.org/3/a0701e/a0701e.pdf. Accessed 11 Mar 2020

FAO (2020) Gateway to poultry production and products. http://www.fao.org/poultry-production-products/production/poultry-species/chickens/en/. Accessed 11 Mar 2020

Farghly MFA, Galal AE, Ahmad EAM (2019) Using wet feed in feeding Japanese quail under summer conditions. Egyptian Poult Sci J 39(2):391–403

Haščík P, Elimam I, Garlík J, Kačániová M, Cubon J, Bobko M, Abdulla H (2012) Impact of beer pollen as feed supplements on the body weight of broiler *Ross 308*. Afr J Biotech 11(89):15596–15599

Hummel D (2015) Climate change, land degradation and migration in Mali and Senegal- some policy implications. Migration and Devt 5(2):211–233

Iheukwumere FC, Herbert U (2003) Physiological responses of broiler chickens to quantitative water restrictions: haematology and serum biochemistry. Int J Poult Sci 2(2):117–119

IPCC (2007) Intergovernmental panel on climate change: AR4-the intergovernmental panel on climate change 4th assessment report. Jackson Institute, University College, London

Kamaruzzaman MA, Chin KY, Ramli ESM (2019) A review of potential beneficial effects of honey on bone health. Evid Based Complementary Altern Med 1–10. https://doi.org/10.1155/2019/8543618

Karaca AG, Parker HM, Yeatman JB, McDaniel CD (2002) The effects of heat stress and sperm quality classification on broiler breeder male fertility and semen ion concentrations. Br Poult Sci 43(4):621–628

Lara LJ, Rostagno MH (2013) Impact of heat stress on poultry production. Animals 3:356–369

Levine S (1962) Plasma-free corticosteroid response to electric shock in rats stimulated in infancy. Science 135:795–796

Liew PK, Zulkifli I, Hair-Bejo M, Omar AR, Israf DA (2003) Effects of early age feed restriction and heat conditioning on heat shock protein 70 expression, resistance to infectious bursal disease and growth in male broiler chickens subjected to heat stress. Poult Sci 82:1879–1885

Manning L, Chadd SA, Baines RN (2007) Water consumption in broiler chicken: a welfare indicator. Worlds Poult Sci J 63:63–71

Mitchell MA, Sandercock DA, Macleod MG, Hunter RR (2005) Thermoregulatory and metabolic heat production responses during acute heat stress in genetically improved broiler chickens. Proceedings of the international poultry scientific forum. Southern Poultry Science Society, Atlanta, p 110

Mustafa YS, Sulehria AQK, Muneer MA, Munir R (2010) Effect of water restriction on the lymphoid organs and production of broilers. Biologia 56(1&2):63–68

Ngaira JKW (2007) Impact of climate change on agriculture in Africa by 2030. Scient Res Essays 2 (7):238–243

Niang I, Ruppel OC, Abdrabo MA, Essel A, Lennard C, Padgham J, Urquhart P (2014) In: Barros VR, Field CB, Dokken DJ, Mastrandrea MD, Mach KJ, Bilir TE, Chatterjee M, Ebi KL, Estrada YO, Genova RC, Girma B, Kissel ES, Levy AN, MacCracken S, Mastrandrea PR, White LL (eds) Climate change: Impacts, adaptation and vulnerability. Par B: regional aspects. Contribution of Working Group II to the Fifth Assessment Report of the Intergovernmental Panel on Climate Change. Cambridge University Press, Cambridge, United Kingdom and New York, pp 1199–1265

Nichelmann M, Tzschentke B (2002) Ontogeny of thermo-regulation in precocial birds. Comp Biochem Physiol 131(4):751–763

Oke OE, Sorungbe FO, Abioja MO, Oyetunji O, Onabajo AO (2016) Effect of different levels of honey on physiological, growth and carcass traits of broiler chickens during dry season. Acta Agric Slov 108:45–53

Olori VE (2008) Breeding broilers for production systems in Africa. Nig Poult Sci J 5(4):173–180

Oryan A, Alemzadeh E, Moshiri A (2016) Biological properties and therapeutic activities of honey in wound healing: a narrative review and meta-analysis. J Tissue Viability 25(2):98–118

Osakwe I, Igwe R (2015) Physiological responses of laying birds fed honey and vitamin C in drinking water. In: A paper presented at conference on international research on food security, natural resource management and rural development, pp 1–6

Prades A, Dornier M, Diop N, Pain JP (2012a) Coconut water uses, composition and properties: a review. Fruits 67:87–107

Prades A, Dornie M, Diop N, Pain JP (2012b) Coconut water preservation and processing: a review. Fruits 67:157–171

Reddy EP, Lakshmi TM (2014) Coconut water- properties, uses, nutritional benefits in health and wealth and in health and disease: a review. J Curr Trends Clin Med Lab Biochem 2(2):6–18

Renaudeau D, Collin A, Yahav S, De Basilio V, Gourdine JL, Collier RJ (2012) Adaptation to hot climate and strategies to alleviate heat stress in livestock production. Animal 6:707–728

Rozenboim I, Tako E, Gal-Garber O, Proudman JA, Uni Z (2007) The effect of heat stress on ovarian function of laying hens. Poult Sci 86(8):1760–1765

Saat M, Singh R, Gamini Sirisinghe R, Nawawi M (2002) Rehydration after exercise with fresh young coconut water, carbohydrate electrolyte beverage and plain water. J Physiol Anthrop Appl Human Sci 21:93–104

Santos RR, Awati A, Roubos-van den Hil PJ, Tersteeg-Zijderveld MH, Koolmees PA, Fink-Gremmels J (2015) Quantitative histo-metric analysis of heat-stress-related damage in the small intestines of broiler chickens. Avian Pathol 44(1):19–22

Shields S, Greger M (2013) Animal welfare and food safety aspects of confining broiler chickens to cages. Animals 3:386–400

Simsek UG, Erisir M, Ciftci M, Tatli Seven P (2014) Effects of cage and flooring housing systems on fattening performance, oxidative stress and carcass defects in broiler chicken. Kafkas Üniversitesi Veteriner Fakültesi Dergisi 20(5):727–733

Syafwan S, Kwakkel RP, Verstegen MWA (2011) Heat stress and feeding strategies in meat-type chickens. Worlds Poult Sci J 67(4):653–674

Tiruneh S, Tegene F (2018) Impacts of climate change on livestock production and productivity and different adaptation strategies in Ethiopia. J Applied Adv Res 3(3):52–58

Tzschentke B (2007) Attainment of thermoregulation as affected by environmental factors. Poult Sci 86:1025–1036

Tzschentke B, Basta D (2002) Early development of neuronal hypothalamic thermosensitivity in birds: influence of epigenetic temperature adaptation. Comp Biochem Physiol 131A:825–832

Tzschentke B, Basta D, Janke O, Maier I (2004) Characteristics of early development of body functions and epigenetic adaptation to the environment in poultry: focused on development of central nervous mechanisms. Avian Poult Biol Rev 15:107–118

Underwood BA (1998) Perspectives from micronutrient malnutrition eradication programs. B World Health Organ 76: 34–37

Wang J, Jin GM, Zheng YM, Li SH, Wang H (2005) Effect of bee pollen on development of immune organ of animal. Zhongguo Zhong Yao Za Zhi 30:1532–1536

Yahav S (2000) Domestic fowl: strategies to confront environmental conditions. Avian Poult Biol Rev 11:81–95

Yahav S, Hurwitz S (1996) Induction of thermotolerance in male broiler chickens by temperature conditioning at an early age. Poult Sci 75:402–406

Yahav S, McMurtry JP (2001) Thermotolerance acquisition in broiler chickens by temperature conditioning early in life – the effect of timing and ambient temperature. Poult Sci 80:1662–1666

Yahav S, Plavnik I (1999) Effect of early-age thermal conditioning and food restriction on performance and thermotolerance of male broiler chickens. Br Poult Sci 40:120–126

Zaid SS, Sulaiman SA, Sirajudeen KN, Othman NH (2011) The effects of Tualang honey on female reproductive organs, tibia bone and hormonal profile in ovariectomised rats-animal model for menopause. BMC Complementary Altern Med 10:82–90

Zulkifli I, Che Norma MT, Israf DA, Omar AR (2000) The effect of early age feed restriction on subsequent response to high environmental temperatures in female broiler chickens. Poult Sci 79:1401–1407

Zulkifli I, Dunnington EA, Gross WB, Siegel PB (1994) Inhibition of adrenal steroidogenesis, food restriction and acclimation to high ambient temperatures in chickens. Br Poult Sci 35:417–426

# 12

# Livestock Breeders' Adaptation to Climate Variability and Change in Morocco's Arid Rangelands

Wadii Snaibi and Abdelhamid Mezrhab

## Contents

Introduction ....
Materials and Methods
Data Collection
Data Analysis
Results and Discussion
Socioeconomic Characteristics of Breeders Surveyed
Analysis of Pastoralists Adaptation Practices in the HPEM
Differentiation of Adaptation Measures According to Breeders' Classes .....
Factors Influencing the Adoption of Climate Change Adaptation Practices
Conclusion
References

**Abstract**

Since the mid-1970s, the high plateaus of eastern Morocco have experienced proven trends of climate change (CC) such as a significant decrease in rainfall amounts and an increase in the droughts' frequency. Consequently, the CC threatens the sustainability of this pastoral ecosystem and negatively affects the

W. Snaibi (✉)
Laboratory Communication, Education, Digital Usage and Creativity, ETIGGE Research Team, Bd. Mohammed VI, University Complex, Mohammed Premier Oujda University, Oujda, Morocco

National Institute of the Agronomic Research of Morocco, CRRA of Oujda, Oujda, Morocco

A. Mezrhab
Laboratory Communication, Education, Digital Usage and Creativity, ETIGGE Research Team, Bd. Mohammed VI, University Complex, Mohammed Premier Oujda University, Oujda, Morocco

breeding of small ruminants, the main local-level livelihood, which becomes more vulnerable due to its high dependence on climatic conditions. This chapter aims to analyze breeders' adaptation practices by taking into account their social stratification based on the size of the sheep flock in possession. Data were analyzed using descriptive statistics, Kruskal-Wallis and Mann-Whitney tests to examine the differences in the adoption' frequency of CC adaptation measures according breeders' classes and Chi-square independence test to identify the factors explaining these observed differences. The analysis of local adaptation practices reveals that they are endogenous but above all curative, aiming at a short-term logic and have a low to medium relevance compared to the specific objective of adaptation to CC. In addition, there are significant differences in the frequency of adoption of CC adaptation strategies (chi-square value = 8.1112, $p = 0.017$, df = 2) within categories of breeders, in particular between small and larger breeders (U statistic = 58.000, $p$ = 0.008). The significant factors explaining these differences are socioeconomic (age, household size, equipment, training, and membership of a basic professional organization). It is therefore recommended to target small breeders as a priority and to set up support measures (equipment, training, funding, organization of breeders).

**Keywords**

Climate change · Adaptation · Pastoralists · Arid rangelands · Morocco

## Introduction

Currently, climate change (CC) has become obvious, due to its manifold effects on human and natural systems, affecting all countries of the world (IPCC 2014). Africa is considered one of the most vulnerable continents to climate change and extreme events (Garcia 2008). This vulnerability is principally attributed to the low level of economic development in its countries, thus generating a weak and limited adaptive capacity to deal with the negative physical, human, and socioeconomic impacts of CC (Besada and Sewankambo 2009; Bruckner 2012).

Within North African countries, Morocco is seen as the most vulnerable to climatic change and extremes, due to the combination of high exposure to climate consequences, marked climate sensitivity (high dependency on rain-fed agriculture, recurrent water stress), and a weak generic adaptive capacity (low income per capita and its unequal distribution) (Yohe et al. 2006; Schilling et al. 2012). Indeed, Morocco has already suffered the impacts of climate change, evidenced by the climate trends observed during the period from 1960 to 2005. Thus, aridity has increased from the south to the north of the country (Mokssit 2012), rainfall had seen a widespread decline (Schilling et al. 2012) of between 3% and 30% (Ezouine and Bouaza 2019), while the temperature has risen (Hulme et al. 2001) from 1.0 to more than 1.8 °C (Morocco 2016). In addition, the frequency, intensity and duration of droughts have increased during the last three decades (Moroccan Meteorological

Office 2007; NIC 2009). Climate change projections for Morocco, exhibit a decrease in annual rainfall amounts of 10–20% by 2100 (Mokssit 2012; Morocco 2016) and an increase in temperature until 2050 between 1.0 °C to 1.2 °C (Schilling et al. 2012) or ranging from 2 °C to 3 °C according to Paeth et al. (2009). Also, the frequency and the duration of droughts are forecasted to rise (Beniston et al. 2007; Bzioui 2012; Schilling et al. 2012). In short, negative climatic changes already observed are likely to continue in the twenty-first century, resulting in warmer and drier conditions (Schilling et al. 2012). Agriculture, the key economic sector and the main provider of jobs, is the most negatively affected by the effects of CC (Morocco 2016). Thereby, rural populations whose rain-fed agriculture is the major source of income are hardest hit because of their high vulnerability to the harmful consequences of climate change (Morocco 2011).

In this context, the study area, namely, the high plateaus of eastern Morocco (HPEM), has experienced proven climate change trends since the late 1970s, such as a substantial decrease in rainfall (Fink et al. 2010; François et al. 2016; Melhaoui et al. 2018) and an increase in temperature and frequency of droughts (Moroccan Meteorological Office 2007; François et al. 2016; Melhaoui et al. 2018). Livestock breeding on rangelands, based mostly on the sheep farming, is the main livelihood and job provider for the local population. This economic activity is vulnerable to CC due to its high dependence on climatic conditions (Bechchari et al. 2014), which are characterized by high intra- and inter-annual variability in rainfall and recurrent droughts (Mahyou et al. 2010; Bechchari et al. 2014). In fact, livestock breeding on arid rangelands is severely affected by the adverse impacts of extreme climate change events since it depends on natural resources and practiced in a fragile and marginal environment such as the pastoral ecosystem of the study area. As highlighted by Hassan (2010) drylands present an intrinsic natural vulnerability generated by a high exposure to significant water stress. Indeed, climate change and extremes in the high plateaus of eastern Morocco threatened the sustainability of pastoral livestock rearing, accentuated the precariousness of the poorest rural households and increased the flow of potential emigrants. Also, extreme climatic events (drought in particular), that have occurred in the HPEM in the past, have caused fodder and water scarcities leading to higher competition for available natural resources and to sometimes brutal pastoral conflicts (Bourbouze and El Aich 2000).

The increase in the occurrence and intensity of climate change-induced droughts and their prolongation over time exacerbated social inequalities among local breeders (Schilling et al. 2012). Indeed, at the time of drought event, small-scale livestock owners face both increased pressure on available pastoral resources and an inability to purchase higher-priced livestock feeds. Thereafter, the size of their herds decreases considerably, while large breeders, their strategy of decapitalization seems to be well under control and their ability to replenish herds is much greater (Bourbouze 2000; Bechchari et al. 2014). In fact, better-off herders are slightly less vulnerable because they are not exclusively or largely dependent on natural resources, have the financial resources to buy livestock feed, and can even take advantage of this opportunity created by the decapitalization of the poorest livestock owners (Kuhn et al. 2010; Bechchari et al. 2014).

In addition, decrease in rainfall and increased droughts' frequency have contributed to the degradation of rangelands in the HPEM, which has also been caused by overgrazing, plowing and anarchic cultivation of marginal areas, uncontrolled land clearing and sedentarization (Mahyou et al. 2010; Maâtougui et al. 2011; Schilling et al. 2012). Furthermore, climate change and extreme weather events, especially, have often limited the success and sustainability of local public interventions in terms of development and poverty reduction.

To deal with or overcome the negative impacts of climate variability and change, livestock herders in the HPEM have undertaken a diversified range of coping and adaptation practices such as pastoral mobility (transhumance), reciprocal grazing agreements with distant pastoral tribes, rearing of mixed species herds, social networks, and intracommunity solidarity to mitigate income shocks (Bourbouze 2000; Bourbouze and El Aich 2000; Schilling et al. 2012). Complementing these traditional coping actions, contemporary adaptation strategies, which are widely implemented in the study area, include: association of cereal crops and livestock farming, breeding of mixed herds of sheep and goats, new form of mobility based on motorization, income diversification and use of emigrants' remittances, commercializing of livestock, storage of livestock feed, using subsidized livestock feed and public programs against drought effects, sale of animals to purchase supplementary livestock feed (Bourbouze 2000; Bourbouze and El Aich 2000; Schilling et al. 2012), and recently the subscription to insurance climatic multi-risks which covers land cultivated with cereals, but not mobile or sedentary livestock rearing.

Nevertheless, these adaptation strategies are in large part of curative scope, low efficient and are less sustainable over time (Bourbouze 2000; Bourbouze and El Aich 2000). In addition, many of these adaptive practices show a relative or low relevance regarding to the specific objective of adapting to climate change and extremes, since they are seen rather as alternative livelihoods more than adaptation actions itself. Thus, they are undertaken by some breeders in order to secure or diversify their livelihoods, for instance, casual work, emigration and the practice of income-generating activities in complement to livestock rearing.

In addition, the adaptive capacity of the herders in the HPEM's area to climate variability and changes depends on the size of the livestock held and the financial and material resources available. So, their adaptive behaviors rely on their respective socioeconomic status (Bechchari et al. 2014). Thereby, small-scale breeders have less options for adaptation, are more severely affected by the observed adverse climatic trends, and are more threatened by the abandonment of livestock rearing (Bourbouze 2000). More globally, Lazarev (2008) pointed out that the main criteria for differentiation between categories of breeders in the study area are socioeconomic characteristics such as the size of the herd exploited and the capital available. Similarly, it is accepted that the susceptibility to the effects of CC and the adaptation capacities differ according to the farmers. In fact, since that the farmers' CC adaptation depends on their respective specific socioeconomic conditions (Below et al. 2012), the design of an effective and appropriate adaptation CC policy and strategies, must take into consideration these differences within human contexts (Fussel 2007) by using local-level analyses (Below et al. 2012).

In view of all of the above, the study is aimed at filling the knowledge gaps relating to the adaptation of breeders in the study area to CC, mainly with regard to the effect of social differentiation within livestock herders on the adoption (implementation) of adaptation strategies to this phenomenon. Concretely, the specific objective is to analyze breeders' practices in adapting to climate variability and change, by taking into account their contrasting socioeconomic conditions mainly the existing differences in the size of the sheep flock in ownership.

## Materials and Methods

### Study Area

Located in the northeast of the country (30S UTM zone), the high plateaus of eastern Morocco (HPEM) are one of the largest pastoral areas in Morocco, covering about 3.5 million hectares. Their soils are generally shallow, low in organic matter and therefore susceptible to wind and water erosion (Mahyou et al. 2016). Water resources are very limited. The HPEM have two gradients going from north to south: the altitude increases regularly from 900 to 1400 m and the climate fluctuates from semi-arid to lower arid and pre-Saharan. Indeed, the climate is of Mediterranean type, but under of a great influence of the Sahara. Average annual rainfall is highly variable, ranging from 143 mm in the south to 201 mm in the north, with respective coefficients of variation of 45% and 34% (Melhaoui et al. 2018). Dry and hot winds which can cause real sandstorms, especially in summer, are frequent. The rangelands of the HPEM are dominated by specific steppe vegetation consisting of steppes at *Stipa tenacissima*, Chamemic steppes at *Artemisia herba alba* and steppes at Chenopodiaceae (*Artrophytum scoparium*). The vast majority of local population derive most of their income from livestock farming, in particular sheep breeding. The herds in the possession of the breeders in the study area are made up of more than two million heads of small ruminants, usually conducted according to an extensive to semi-extensive rearing system. The HPEM area can be seen as a suitable and representative site for better understanding of adaptation to CC in Morocco's pastoral ecosystems.

### Data Collection

Data collection method consisted of a literature review and a survey of 167 breeders, heads of pastoralist households. Relevant literature available from local extension and agricultural development agencies were consulted to acquire a complete and clear overview about the CC issue in the HPEM's area, mainly with regard to livestock practices and core endogenous adaptation measures implemented by the breeders in response to climate variability and change. The survey of herders focused on the socioeconomic characteristics of households and the adaptation practices embraced to reduce the effects of perceived CC. The basic study unit is the pastoralist household, as at this level, decisions relating to adaptation to climate variability and change are taken (Below et al. 2012).

Given that sheep farming is the main activity of the pastoralist households in the study area, the size of the sheep flock in ownership was chosen as the criterion of discrimination between herders. Thereby, three classes of livestock breeders have been identified and this in agreement with local agricultural extension agencies. Large breeders are those with a sheep herd exceeding 300 heads, medium breeders with own sheep flocks of between 101 and 300 heads and the small livestock owners with the number of sheep in possession is less than or equal to 100 heads. Based on the respective representativeness of these three breeders' categories in the study area (Bechchari et al. 2014), respondent herders were randomly selected. The distribution of breeders surveyed by class is as follows: 96 small, 47 medium, and 24 large breeders to give a total of 167 livestock producers.

## Data Analysis

Data collected on the adaptation practices implemented by the livestock herders in the study area were analyzed using descriptive statistics and the tests of Kruskal-Wallis, Mann-Whitney U and of Chi-square independence. Descriptive statistics have been made, with regard to the socioeconomic characteristics of breeders surveyed and the different local adaptation measures. The Chi-square independence test was used to highlight the relationship that may exist between the developed endogenous practices and the categories of herders (small, medium, large). In addition, the strength of this possible statistical association was measured using Cramer's V. coefficient. This test was also carried out to verify whether the observed differences within the three breeders' classes is linked to their respective socioeconomic conditions at household level. Furthermore, Kruskal-Wallis test is an appropriate nonparametric test for comparing more than two independent samples. It is a rank-based test which can be used to test whether such samples come from the same distribution (Ostertagova et al. 2014). The null hypothesis is the following: all the populations have the same median or no significant difference between the groups (samples). In our case, the Kruskal-Wallis test was used to testing for difference in the frequency of adoption of adaptation strategies among three livestock breeders' classes. If the Kruskal-Wallis statistic is significant, a nonparametric multiple comparison method (Mann-Whitney U test) is used to find out which classes of herders are different from the others. Also, using the Mann-Whitney U test, we assessed if the adoption of adaptation practices to CC differed significantly between the three classes of breeders which are based on the size of the owned sheep herd. Indeed, this test was realized to show if there is a significant difference between the "adoptive" and "non-adoptive" groups according to the number of sheep in possession. Generally, the Mann-Whitney U test is a non-parametric statistical test used to determine if two independent samples come from the same population or from 2 different populations. Indeed, this test was used because the assumptions of use of the Student's t test were not verified (non-comparable variances and non-normal distribution of the dependent variable).

## Results and Discussion

### Socioeconomic Characteristics of Breeders Surveyed

The average age of breeders is 52 years with almost 60% of them who have an age greater than or equal to 50 years. Livestock breeders are in large part without level of education (70%). The rearing of small ruminants is the main economic activity for 83% of herders sampled. Approximately 35% of respondents are engaged in an ancillary activity (mainly the temporary employment: 26%). The size of households is an average of 8 people. The average number of labor force by household is around 2 persons. The employ of shepherds is observed in 36% of cases. The rate of emigration is low and is close to 17%. The types of dominant habitat are the tents and concrete houses with respectively 36% and 24%. The rate of connection to the public electricity network is low, either almost 20%. The distance to the nearest market is relatively long and is 60 km.

The majority of the breeders (83%) had agricultural land, but only 16% of the herders surveyed owned an irrigated plot. In effect, on average, arable land was about 31 ha and irrigated area did not exceed the 0.37 ha. Breeders are moderately equipped, since more than 61% of them possesses at least one agricultural equipment and/or transport (possession trucks: 27%, tractors: 14% and water tanks: 50%). The average number of sheep and goats are respectively of 166 and 33 heads/breeder. More than 52% of the herders have at least one head of bovine cattle.

Access to formal credit is low not exceeding 23%.

### Analysis of Pastoralists Adaptation Practices in the HPEM

Table 1 showed that the breeders in the HPEM have implemented, over time, a wide and diversified range of adaptation practices to cope with the harmful effects of hazards and climate changes.

These adaptation and coping measures are almost all of endogenous origin, thus attesting on a great intrinsic capacity for adaptation based on the accumulation of experiences and initiatives of the breeders in this arid pastoral ecosystem where climatic conditions are difficult and unpredictable. Local adaptation practices to climate variability and change can be grouped into two main categories: (1) Adjustment of farm management and pastoral practices and (2) Partial shift to alternative livelihood options. A large part of the adaptation actions identified concerns the first dimension.

Contemporary adaptive strategies in the study area are mostly individualized. This testifies to the spectacular rise of an individualism of spirit and action and of refocusing on the individual to the detriment of the social group to which he belongs (Bourbouze and El Aich 2000). These authors added that it is now at the individual level that one must know how to protect oneself from environmental and economic risks and no longer at the community level. Whereas during the first half of the last century, local customary institutions organized the access and use of collective

**Table 1** Distribution of herders surveyed according to adaptive measures implemented (in %)

| Adaptation practices | Freq. | Perc. | Dur. | Inten. | Adeq. |
|---|---|---|---|---|---|
| (1) Adjustment of farm management and pastoral practices | | | | | |
| Mixed livestock crop farming system | 139 | 83.2 | LT | P | M |
| Profit of state agricultural programs | 132 | 79 | ST | C | M |
| Diversification of livestock species | 130 | 77.8 | LT | P | M |
| Climate multi hazard insurance | 80 | 47.9 | ST | C | H |
| Storage of animal feed | 80 | 47.9 | ST | C | H |
| Herd mobility | 67 | 40.1 | ST | C | H |
| Regular Sale of animals to stock up on feed | 79 | 47.3 | ST | C | L |
| Sale of the animal in a good physical state | 73 | 43.7 | LT | P | M |
| Practice of fattening | 51 | 30.5 | ST | C | M |
| Credit from speculators livestock feed | 37 | 22.2 | ST | C | L |
| Privative appropriation of rangelands | 30 | 18 | ST | C | M |
| Irrigated agriculture and livestock integration | 22 | 13.2 | LT | P | H |
| (2) Partial shift to alternative livelihood options | | | | | |
| Conversion of livestock capital into land capital | 67 | 40.1 | LT | P | L |
| Casual labor | 55 | 32.9 | ST | C | L |
| Internal or external emigration in search of jobs | 28 | 16.8 | LT | P | L |
| Collection of truffles as additional income | 14 | 8.4 | ST | C | L |

Freq.: Frequency. Perc.: Percentage. Duration: ST (Short term); LT (Long term). Intention: C (Curative action); P (Preventive action). Adequacy: H: High; M: Medium; L: Low

pastoral resources and guaranteed intra- and inter-tribal solidarity for the survival of their ethnic groups in times of climate crisis. The decline of these traditional structures for several decades has been caused mainly by a public policy which has fragmented tribal organizations in favor of modern administrative and elected institutions and encouraged sedentarization (Rachik 2007). Negative practices of breeders, especially the large and influential among them, such as the cultivation of rangelands and their appropriation for private use, have also contributed to the weakening of these tribal structures (Bourbouze 2000; Bechchari et al. 2014).

In addition, the majority of the adaptation measures adopted are part of a short-term temporal perspective, thus allowing breeders to buffer climate risks and reduce their negative consequences through the practices of curative type (for example the regular sale of animals to purchase livestock feed, benefit from public interventions such as subsidized fodder). This testifies to the predominance of short-term and reactive vision of adaptation at local level. Furthermore, most local adaptive practices show a low to medium adequacy in relation to the main objective of adaptation to CC, as their purpose is not a specific response to this phenomenon. They are embraced by breeders either to improve their rearing productivity and, thus, contribute to reducing their vulnerability to hazards and hostile climatic changes, such as diversification of livestock species and practice of fattening (medium adequacy) or to increase their income, meet the financial needs of their farms and maintain their economic activity, e.g., regular sale of animals to purchase livestock feed, credit from resellers of livestock feed and casual employment (low adequacy). Only four adaptation measures which were highly relevant in relation to climate change. They are: transhumance or herd mobility, storage of livestock feeds, subscription to climate insurance and the integration of livestock breeding and irrigated agriculture. However, these adaptive actions depend on the socioeconomic status of the breeder (mostly implemented by large herders) or they are very spatially localized like irrigated agriculture.

In line with our findings, Bourbouze (2000) and Bourbouze and El Aich (2000) highlighted that most adaptation practices to variability and climatic risks (drought in particular) implemented by the breeders in the HPEM are of curative type and provide only relative protection. The only really effective strategies against climatic hazards were the cereal-livestock association, livestock feed supplements purchased through animal sales, remittances of emigrants and irrigated agriculture by pumping, but even these are little effective in a sustainable way. Bechchari et al. (2014) emphasized that the adaptive capacity of herders in the HPEM's area with regard to climate-related risks and changes, closely depends on their respective socioeconomic characteristics. In fact, the socioeconomic status of breeders in the study area largely affects their drought adaptation responses and more generally their pastoral practices and their way of using the rangelands (Mahdi 2007; Lazarev 2008).

## Differentiation of Adaptation Measures According to Breeders' Classes

The Table 2 showed general divergence in the frequencies of adaptation measures implemented within the categories of breeders (small, medium and large) which are based on the size of the sheep flock in ownership. Thus, large livestock owners adopt with higher frequencies, most of the endogenous adaptation practices in response to perceived climate variability and change, compared to the two other categories. In addition, practices of the first dimension of adaptation measures are more frequently implemented by large breeders. They include strategies for the diversification of the productions (integration of livestock breeding and cereal farming, breeding of mixed flocks of sheep and goats), improvement of rearing productivity (practice of the fattening, selection and reproduction of powerful races), transhumance and market orientation (finished products of good taste quality). In fact, all of these adaptation strategies require considerable financial resources, which manifests itself in much lower adoption frequencies among small and medium-sized breeders.

By relying on their social status and their relationship networks, large breeders adopt an opportunistic land strategy by constantly conquering vast new areas of collective rangelands for their private use. As evidenced by the significant differences between the average available agricultural areas, which are 86, 37 and 15 hectares respectively for great, medium and small breeders. Recently, large herders and some medium breeders have massively subscribed to climatic multi-risk insurance. The adoption of this practice has two objectives. First, climate insurance contracts represent, in the eyes of local breeders, justifications approving the legitimacy of the ownership of conquered rangelands in order to annex them definitively

**Table 2** Frequencies of adaptation measures in percentage according to breeders' classes and results of the Chi-square independence test

| Adaptation practices | Small (n = 96) | Medium (n = 47) | Large (n = 24) | $\chi^2$ Pearson | Cramer's V |
|---|---|---|---|---|---|
| Mixed livestock crop farming system[a] | 74 | 93.6 | 100 | 14.384 (0.001) | 0.293[M] |
| Profit of state agricultural programs | 75 | 83 | 87.5 | 2.423 (0.298) | 0.120[L] |
| Diversification of livestock species | 71.9 | 85.1 | 87.5 | 4.718 (0.095) | 0.168[L] |
| Climate multi hazard insurance[a] | 28.1 | 70.2 | 83.3 | 36.493 (0.000) | 0.467[S] |
| Storage of animal feed[a] | 36.5 | 63.8 | 62.5 | 11.865 (0.003) | 0.267 [M] |
| Herd mobility[a] | 27.1 | 48.9 | 75 | 20.466 (0.000) | 0.350[M] |
| Regular Sale of animals to stock up on feed | 44.8 | 53.2 | 45.8 | 0.917 (0.632) | 0.074[L] |
| Sale of the animal in a good physical state[a] | 33.3 | 48.9 | 75 | 14.273 (0.001) | 0.292[M] |
| Practice of fattening[a] | 20.8 | 31.9 | 66.7 | 19.072 (0.000) | 0.338[M] |
| Credit from speculators livestock feed[a] | 18.8 | 19.1 | 41.7 | 6.189 (0.045) | 0.193[L] |
| Privative appropriation of rangelands[a] | 13.5 | 17 | 37.5 | 7.518 (0.023) | 0.212[M] |
| Irrigated agriculture and livestock integration | 8.3 | 21.3 | 16.7 | 4.920 (0.085) | 0.172[L] |
| Conversion of livestock capital into land capital[a] | 29.2 | 44.7 | 75 | 17.355 (0.000) | 0.322[M] |
| Casual labor[a] | 42.7 | 27.7 | 4.2 | 13.736 (0.001) | 0.287[M] |
| Internal or external emigration in search of jobs | 13.5 | 21.3 | 20.8 | 1.685 (0.431) | 0.100[L] |
| Collection of truffles as additional income[a] | 13.5 | 2.1 | 0 | 7.917 (0.019) | 0.218[M] |

Note: Cramer's V. value: S: Strong (between 0.40 and 0.80); M: Moderate (between 0.20 and 0.40); L: Low (between 0.10 and 0.20). Values in parentheses in $\chi^2$ Pearson represent the asymptotic significance (bilateral)

[a]Significant at 5% level

later. Also, this measure (insurance) allows subscribers significant financial compensation following of climatic hazards (drought in particular) equals to 600 MAD per hectare.

Large livestock owners have benefited from contracts of sale with supermarkets of many urban agglomerations in the north of the country during the feasts of sacrifice, thanks to the USAID' initiative. Thus, they were able to achieve very interesting butcher performances (well-finished animal products with appreciable gustative qualities). This explains the strong orientation of their breeding activity towards the market. In addition, thanks to their accumulated savings, large livestock keepers invest more and more in activities of speculation, including the real estate, in order to overcome the unfavorable climatic and economic conditions affecting their livestock rearing activity.

As for small herders, they frequently shift to additional off-farm livelihoods to complete or diversify their income sources. Thus, they are forced to engage in other small-scale activities in addition to the livestock breeding, such as temporary labor, collecting truffles and small trades, in order to satisfy both the needs of their families and those of their meager herds especially in the event of climatic vagaries (drought) prolonged in time. The poorest of them, after the sale of all their herds, find themselves decapitalized. In the absence of support from relatives or the State, they opt for rural exodus as the final solution.

In addition, the results of the Chi-square independence test revealed that there was a statistically significant relationship between most adaptation practices (11/16 measures identified) and livestock breeders' classes (Table 2). The Cramer's V. coefficient measuring the strength of this statistical relationship has medium to high values. This association of significant magnitude indicates, thereby, that the larger the size of the sheep herd in possession, the higher the frequency of adoption (implementation) of adaptation practices.

As highlighted by Bourbouze (2000), the choices for small breeders regarding adaptation to adverse impacts of climate variability and extremes are much narrower and their most common strategy is to regularly sell animals in the souk so that they can buy livestock feed, water their herds and feed their families. Thereafter, either they abandon livestock rearing after having sold all their flocks, or they migrate to the nearest urban centers in search of small trades. Whereas the decapitalization strategy seems well controlled among large breeders (Bourbouze and El Aich 2000), they can even take advantage of periods of climatic crisis to increase the size of their herds to the detriment of the poorest breeders (Schilling et al. 2012). Bechchari et al. (2014) underlined that the portfolio of strategies for adapting to variability and climate change implemented by large breeders is much denser and more diversified. It includes the acquisition of equipment (trucks, tractors, water tanks), the appropriation of large areas of rangelands, the profit of the best grazing sites and the commercial speculation in livestock in order to make a good profit on their invested capitals. In contrast, the possibilities offered for small breeders, are very limited and mainly boil down to the regular sale of their herds, the search for additional activities to breeding and rural exodus.

The Kruskal-Wallis test was conducted to examine the differences in the frequency of adoption of climate change adaptation practices according to the classes of breeders. Significant differences (chi square $= 8.1123$, $p = 0.017$, df $= 2$) were found between the three categories of herders. In addition, the distribution of the majority of adaptive measures differs significantly between classes of breeders (Table 3).

After the Kruskal-Wallis test led us to reject the null hypothesis of similarity between breeders' categories in terms of frequency of adoption of adaptation and coping actions, we performed pairwise comparisons using the Mann-Whitney test to

**Table 3** Results of the Kruskal-Wallis test

| Adaptation practices | Chi-square value | Sig-p |
|---|---|---|
| Mixed livestock crop farming system** | 14.298 | 0.001 |
| Profit of state agricultural programs | 2.408 | 0.300 |
| Diversification of livestock species | 4.690 | 0.096 |
| Climate multi hazard insurance*** | 36.275 | 0.000 |
| Storage of animal feed** | 11.794 | 0.003 |
| Herd mobility*** | 20.344 | 0.000 |
| Regular Sale of animals to stock up on feed | 0.912 | 0.634 |
| Sale of the animal in a good physical state** | 14.187 | 0.001 |
| Practice of fattening*** | 18.958 | 0.000 |
| Credit from speculators livestock feed* | 6.152 | 0.046 |
| Privative appropriation of rangelands* | 7.473 | 0.024 |
| Irrigated agriculture and livestock integration | 4.891 | 0.087 |
| Conversion of livestock capital into land capital*** | 17.251 | 0.000 |
| Casual labor** | 13.654 | 0.001 |
| Internal or external emigration in search of jobs | 1.675 | 0.433 |
| Collection of truffles as additional income* | 7.869 | 0.020 |

Note: Significance: ***$P < 0.01$, **$P < 0.05$, *$P < 0.1$; Degree of freedom = 2

determine which classes of breeders are different. The results of this test show that there is a highly significant difference (U statistic = 58.000, $p = 0.008$) between small and large breeders in adopting CC adaptation practices. The other differences between on the one hand small and medium breeders (U = 92.000, $p = 0.174$) and between medium and large herders on the other hand (U = 79.500, $p = 0.067$) were found not to be significant. This shows that breeders' adaptation in the study area depends mainly on their social status or economic power, expressed by the size of the sheep herd in possession.

## Factors Influencing the Adoption of Climate Change Adaptation Practices

In order to confirm this relationship between the adaptation measures to climate variability and change implemented by the breeders in the study area and the size of the sheep herd in possession, we carried out the Mann-Whitney U test. This test is often employed to compare differences between two independent groups when the dependent variable is continuous, but not normally distributed. In our case, the two independent groups are the "adopters" and "non-adopters" of CC adaptation practices, while the continuous dependent variable is the sheep herd size. The mean rank column shows mean rank for the two groups tested (adopters and non-adopters groups). This column is very useful because it indicates which group can be considered as having the higher size of sheep herd, overall; namely, the group with the highest mean rank. In this case, the adopter group had the highest size of sheep herd. Indeed, the first group (adopters) have higher mean ranks than those of non-adopters, except for the case of two coping actions, namely, "Casual labor" and "Collection of truffles as additional income" (Table 4). In fact, these two practices are usually pursued either as a complement to a breeding activity that does not allow the household to meet its needs or in the event of abandonment of livestock farming (decapitalization frequently caused by successions of prolonged episodes of drought). They are practiced mainly by the small breeders with respectively 75 and 93%.

Furthermore, the Test Statistics column shows us the actual significance value of the Mann-Whitney U test. Specifically, this column provides the test statistic, U statistic, as well as the asymptotic significance p-value. From our data, it can be

**Table 4** Mann-Whitney U test showing the relationship between sheep herd size and adaptive practices

| | Mean rank | | | |
|---|---|---|---|---|
| Adaptation practices | Yes | No | (U, *p*) | Z |
| Mixed livestock crop farming system | 92.64 | 41.11 | (745, 0.000) | −5.149 |
| Profit of state agricultural programs | 87.16 | 72.07 | (1892.5, 0.100) | −1.643 |
| Diversification of livestock species (sheep & goats) | 90.33 | 61.76 | (1582, 0.002) | −3.174 |
| Climate multi hazard insurance | 105.35 | 64.37 | (1772, 0.000) | −5.476 |
| Storage of animal feed | 96.01 | 72.96 | (2519.5, 0.002) | −3.079 |
| Herd mobility | 107.60 | 68.19 | (1768.5, 0.000) | −5.168 |
| Regular Sale of animals to stock up on feed | 91.88 | 76.93 | (2853.5, 0.046) | −1.997 |
| Sale of the animal in a good physical state | 100.18 | 71.44 | (2250, 0.000) | −3.813 |
| Practice of fattening | 105.96 | 74.34 | (1838, 0.000) | −3.895 |
| Credit from speculators livestock feed | 98.43 | 79.89 | (1871, 0.039) | −2.059 |
| Privative appropriation of rangelands | 106.17 | 79.15 | (1390, 0.006) | −2.774 |
| Irrigated agriculture and livestock integration | 97.68 | 81.92 | (1294, 0.154) | −1.425 |
| Conversion of livestock capital into land capital | 101.54 | 72.25 | (2174.5, 0.000) | −3.841 |
| Casual labor | 59.39 | 96.08 | (1726.5, 0.000) | −4.612 |
| Internal or external emigration in search of jobs | 94.36 | 81.91 | (1656, 0.214) | −1.243 |
| Collection of truffles as additional income | 27.46 | 89.17 | (279.5, 0.000) | −4.574 |

Note: U: U statistic of Mann-Whitney test, *p*: the value of p of the test

concluded that the size of the sheep herd in the adopter group was significantly higher than in the non-adopter group for the majority of adaptation strategies practiced by the livestock owners in the study area.

In line with our findings, herd size influences positively and significatively the likelihood that pastoralists implement adaptation strategies in the face of climate change (Balew et al. 2014; Berhanu and Beyene 2015; Opiyo et al. 2015).

Other socioeconomic characteristics at household-level, using the category of the breeder as a classification variable, could explain significant observed differences in the frequency of climate change adaptation practice's adoption (implementation) within breeders' classes (small, medium, large). The results of chi-square independence test contained in the Table 5 show that there is a significant relationship between the category of herder and the following factors: age of breeder ($\chi^2 = 19.020, p = 0.000$), household size ($\chi^2 = 30.720, p = 0.000$), no practice of ancillary activity ($\chi^2 = 25.662, p = 0.000$), employment of shepherds ($\chi^2 = 42.124$, $p = 0.000$), land ownership ($\chi^2 = 14.384$, $p = 0.001$), possession of truck ($\chi^2 = 17.754, p = 0.000$), possession of tractor ($\chi^2 = 49.740, p = 0.000$), possession of water tank ($\chi^2 = 33.902, p = 0.000$), possession of a motor pump ($\chi^2 = 39.048$, $p = 0.000$), veterinary care ($\chi^2 = 16.005, p = 0.000$), training received ($\chi^2 = 7.850$, $p = 0.020$) and membership to a technical supervisory structure, namely, National Association of sheep and goat breeders- ANOC ($\chi^2 = 27.454, p = 0.000$).

Large livestock producers have a higher average age, 62 years, compared to 49 years for small breeders (Table 6). They are ones who benefited the most from training actions in livestock rearing, development and management of rangelands and other technical topics of interest (42% vs. 17%), adhered massively to the breeders' organization, namely, ANOC (58% vs. 11%). These elements indicate that large herders have accumulated a great pastoralism experience compared to small-scale breeders. Piya et al. (2013) and Tiwari et al. (2014) pointed out that the training received improved the adaptive capacity of farmers in the face of climate change. As highlighted by Tiwari et al. (2014) and Taruvinga et al. (2016), membership in the community-based organizations increases the adoption of climate change coping strategies. Yila and Resurreccion (2013) and Mabe et al. (2014) underlined that farming experience significantly and positively influences the implementation of CC measures, respectively in the semiarid Nguru Local Government Area, Northeastern Nigeria, and in Northern Ghana. In addition, large breeders are far largely more endowed with production factors than small breeders such as labor force (4 vs.. 1 people), employment of shepherds (88 vs. 19%), land size (86 vs. 15 ha), small ruminants herd size (684 vs. 72 heads), bovine cattle flock size (7 vs. 1 head), number of equipment owned (5 vs. 1) and veterinary care (83 vs. 45%). In line with our findings, Yila and Resurreccion (2013) pointed out that labor force was a significant determinant for farmers' adaptation practices to climate change in the semiarid Nguru Local Government Area, Northeastern Nigeria. Debalke (2011) and Ndamani and Watanabe (2016) highlighted that land size was a determinant factor that influences farmers' climate change coping strategies, respectively, in north shoa zone of Amhara region Ethiopia and Lawra district of Ghana. Berhanu and Beyene (2015) and Opiyo et al. (2015) have found that herd size affected positively and

**Table 5** Significant factors (categorical variables) explaining the differences observed in the frequency of adoption of adaptation practices between categories of breeders

| Categorical variables | Small (%) | Medium (%) | Large (%) | Chi-square value | Sig-p |
|---|---|---|---|---|---|
| Age | | | | 19.020 | 0.000 |
| <50 years | 54 | 32 | 8 | | |
| >=50 years | 46 | 68 | 92 | | |
| Ancillary activity | | | | 25.662 | 0.000 |
| Yes | 47 | 28 | 8 | | |
| No | 53 | 72 | 92 | | |
| Household size | | | | 30.720 | 0.000 |
| <8 people | 70 | 38 | 12 | | |
| >=8 people | 30 | 62 | 88 | | |
| Employment of shepherds | | | | 42.124 | 0.000 |
| Yes | 19 | 47 | 88 | | |
| No | 81 | 53 | 12 | | |
| Land ownership | | | | 14.384 | 0.001 |
| Yes | 74 | 94 | 100 | | |
| No | 26 | 6 | 0 | | |
| Favorable pastures | | | | 12.160 | 0.002 |
| Yes | 14 | 36 | 38 | | |
| No | 86 | 64 | 62 | | |
| Equipment | | | | 20.399 | 0.000 |
| Yes | 50 | 64 | 100 | | |
| No | 50 | 36 | 0 | | |
| Possession of truck | | | | 17.754 | 0.000 |
| Yes | 17 | 32 | 58 | | |
| No | 83 | 68 | 42 | | |
| Possession of tractor | | | | 49.740 | 0.000 |
| Yes | 2 | 17 | 58 | | |
| No | 98 | 83 | 42 | | |
| Possession of cart | | | | 9.995 | 0.007 |
| Yes | 21 | 2 | 8 | | |
| No | 79 | 98 | 92 | | |
| Possession of tank | | | | 33.902 | 0.000 |
| Yes | 34 | 55 | 100 | | |
| No | 66 | 45 | 0 | | |
| Possession of pump | | | | 39.048 | 0.000 |
| Yes | 12 | 30 | 75 | | |
| No | 88 | 70 | 25 | | |
| Veterinary care | | | | 16.005 | 0.000 |
| Yes | 45 | 70 | 83 | | |
| No | 55 | 30 | 17 | | |
| Training received | | | | 7.850 | 0.020 |

(continued)

**Table 5** (continued)

| Categorical variables | Small (%) | Medium (%) | Large (%) | Chi-square value | Sig-p |
|---|---|---|---|---|---|
| Yes | 17 | 17 | 42 | | |
| No | 83 | 83 | 58 | | |
| BO Membership | | | | 27.454 | 0.000 |
| Yes | 11 | 38 | 58 | | |
| No | 89 | 62 | 42 | | |

Note: BO: Breeder Organization

**Table 6** Significant factors (continuous variables) explaining the differences observed in the frequency of adaptation practices' adoption between categories of breeders

| Variable | Category | Minimum | Maximum | Mean | Standard deviation |
|---|---|---|---|---|---|
| Age | Small | 22 | 85 | 49 | 13.94 |
| | Medium | 24 | 79 | 53 | 12.66 |
| | Large | 44 | 79 | 62 | 8.60 |
| Household size | Small | 1 | 14 | 6 | 2.63 |
| | Medium | 3 | 16 | 9 | 3.00 |
| | Large | 5 | 23 | 12 | 4.47 |
| Labor force | Small | 0 | 3 | 1 | 0.93 |
| | Medium | 0 | 6 | 2 | 1.47 |
| | Large | 0 | 12 | 4 | 2.32 |
| Land size | Small | 0 | 52 | 15 | 14.81 |
| | Medium | 0 | 196 | 37 | 41.05 |
| | Large | 8 | 300 | 86 | 82.68 |
| Equipment | Small | 0 | 4 | 1 | 1.19 |
| | Medium | 0 | 6 | 2 | 1.77 |
| | Large | 2 | 14 | 5 | 2.96 |
| Small ruminants herd size | Small | 0 | 180 | 72 | 37.31 |
| | Medium | 101 | 500 | 214 | 78.57 |
| | Large | 355 | 1400 | 684 | 260.00 |
| Bovine cattle herd size | Small | 0 | 10 | 1 | 2.20 |
| | Medium | 0 | 13 | 3 | 3.70 |
| | Large | 0 | 50 | 7 | 10.38 |

significantly the probability that pastoralists put in place CC adaptation measures. Hassan and Nhemachena (2008) and Ouédraogo et al. (2010) stressed that ownership of heavy machinery or agricultural equipment improves the coping capacities of farmers to deal with the negative effects of climate variability and extremes.

Furthermore, large breeders have specialized over time in the extensive breeding of small ruminants, while small-scale herders have been forced to practice non-farm activities such as casual labor, small trades, and the collection of truffles to obtain additional income necessary for their survival. In fact, almost all large livestock owners (92%) do not carry out any ancillary activity generating additional income, unlike 53% for small herders. Given that the occupation of the farmer was an indication of the total amount of time available for farming activities (Gbetibouo 2009), the off-farm employment may present a constraint to the adoption of technology because it consumes time which must be devoted to the management of agricultural activities (McNamara et al. 1991). Also, large breeders are the ones who benefit most from the best grazing sites (38 vs. 14%). Berhanu and Beyene (2015) expressed that the traditional pastoralism represents a resilient and unique system of adaptation to hostile and unpredictable climatic variability in dryland ecosystems.

In accordance with the results presented above, several previous studies showed that the significant factors influencing African farmers' adaptation to climate change are socioeconomic household-level variables, and this in many countries such as Ethiopia (Balew et al. 2014; Berhanu and Beyene 2015), South Africa (Taruvinga et al. 2016), Ghana (Ndamani and Watanabe 2016), Tanzania (Below et al. 2012), Nigeria (Obayelu et al. 2014), Kenya (Opiyo et al. 2015), and Uganda (Nabikolo et al. 2012).

## Conclusion

In response to perceived climate changes and extremes, livestock breeders in the high plateaus of eastern Morocco have implemented differently many adaptation practices, depending on their respective socioeconomic conditions at household level. Thereby, contrasting socioeconomic characteristics, mainly the size of sheep flock in ownership, allowed large breeders to adopt (put in place) with higher frequency most of adaptive measures compared to small-scale herders who have much more limited possibilities or choices. This social inequality could be exacerbated in the future due to adverse climatic trends (increase in the frequency and intensity of droughts, reduction in rainfall amounts) caused by current and future climate change, in the absence of support measures targeting primarily small breeders who are the most vulnerable group to this climatic phenomenon.

In addition, future studies relating to climate change adaptation in the study area should first investigate the perceptions of livestock producers. Indeed, local perceptions toward climate change influence farmers' decisions on whether or not to adapt (Deressa et al. 2009), are useful for the development of relevant and appropriate adaptation policies and strategies (Opiyo et al. 2015), and are also an important factor influencing the success of the adaptation actions to be implemented (Tesfahunegn et al. 2016).

## References

Balew S, Agwata J, Anyango S (2014) Determinants of adoption choices of climate change adaptation strategies in crop production by small scale farmers in some regions of Central Ethiopia. J Nat Sci Res 4(4):78–93

Bechchari A, El Aich A, Mahyou H, Baghdad B, Bendaou M (2014) Etude de la dégradation des pâturages steppiques dans les communes de Maâtarka et Béni Mathar (Maroc oriental). J Mater Environ Sci 5(S2):2572–2583

Below TB, Mutabazi KD, Kirschke D, Franke C, Sieber S, Siebert R, Tscherning K (2012) Can farmers' adaptation to climate change be explained by socioeconomic household-level variables? Glob Environ Chang 22(1):223–235. https://doi.org/10.1016/j.gloenvcha.2011.11.012

Beniston M, Stephenson D, Christensen O et al (2007) Future extreme events in European climate: an exploration of regional climate model projections. Clim Chang 81:71–95

Berhanu W, Beyene F (2015) Climate variability and household adaptation strategies in Southern Ethiopia. Sustainability 2015(7):6353–6375. https://doi.org/10.3390/su7066353

Besada H, Sewankambo N (2009) Climate change in Africa: adaptation, mitigation and governance challenges. CIGI Special Report. Centre for International Governance Innovation = Centre pour l'innovation dans la gouvernance internationale, Waterloo, p 49

Bourbouze A (2000) Pastoralisme au Maghreb: la révolution silencieuse. Fourrages 161:3–21

Bourbouze A, El Aich A (2000) Gestion des parcours et des troupeaux en régions steppiques et réponse à l'aléa climatique. Dans: Guessous F, Rihani N, Ilham A (eds) Livestock production and climatic uncertainty in the Mediterranean. EAAP Publication no 94. Wageningen Pers, Wageningen, pp 307–319

Bruckner M (2012) Climate change vulnerability and the identification of least developed countries (LDCs). The United Nations Development Policy and Analysis Division Department of Economic and Social Affairs. CDP Background Paper no 15, p 29. https://www.un.org/development/desa/dpad/wp-content/uploads/sites/45/publication/CDP-bp-2012-15.pdf

Bzioui M (2012) Gestion des ressources en eau au Maroc et changement climatique. In: Ellinor ZM, Bennis A (eds) Environnement et changement climatique au Maroc- Diagnostic et Perspectives-. Konrad-Adenauer-Stiftung e.V., Bureau au Maroc, Rabat, pp 63–69

Debalke NM (2011) Determinants of farmers' preference for adaptation strategies to climate change: evidence from north shoa zone of Amhara region Ethiopia. MPRA Paper no 48753. Ambo University, Addis Ababa University

Deressa TT, Hassan RM, Ringler C, Alemu T, Yesuf M (2009) Determinants of farmers' choice of adaptation methods to climate change in the Nile Basin of Ethiopia. Glob Environ Chang 19:248–255

Ezouine D, Bouaza N (2019) Climate finance strategy in Morocco. Am J Clim Chang 8:482–501. https://doi.org/10.4236/ajcc.2019.84026

Fink A, Christoph M, Born K et al (2010) Climate. In: Speth P, Christoph M, Diekkrüger B (eds) Impacts of global change on the hydrological cycle in west and Northwest Africa. Springer, Heidelberg, pp 54–58

François A, Gauché E, Génin A (2016) L'adaptation des territoires aux changements climatiques dans l'Oriental marocain: la vulnérabilité entre action et perceptions. VertigO 16(1):1–25. https://doi.org/10.4000/vertigo.17177. http://journals.openedition.org/vertigo/17177

Fussel HM (2007) Adaptation planning for climate change: concepts, assessment approaches, and key lessons. Sustain Sci 2(2):265–275

Garcia D (2008) The climate security divide: bridging human and national security in Africa. Afr Secur Rev 17(3):1–17. https://doi.org/10.1080/10246029.2008.9627480

Gbetibouo GA (2009) Understanding farmers' perceptions and adaptations to climate change and variability – the case of the Limpopo Basin, South Africa. IFPRI Discussion Paper 00849. Centre for Environmental Economics and Policy in Africa, University of Pretoria. http://www.ifpri.org/publication/understanding-farmers-perceptions-and-adaptations-climate-change-and-variability

Hassan RM (2010) Implications of climate change for agricultural sector performance in Africa: policy challenges and research agenda. J Afr Econ 19:77–105. https://doi.org/10.1093/jae/ejp026

Hassan RM, Nhemachena C (2008) Determinants of African farmers strategies for adapting to climate change: multinomial choice analysis. Afr J Agric Resour Econ 2(1):83–104

Hulme M, Doherty R, Ngara T, New M, Lister D (2001) African climate change: 1900–2100. Clim Res 17(2):145–168

International Panel on Climate Change (IPCC) (2014) Synthesis report. In: Pachauri RK, Meyer LA (eds) Climate change 2014: contribution of working groups I, II and III to the fifth assessment report of the intergovernmental panel on climate change. Core writing team. IPCC, Geneva. 151pp. https://www.ipcc.ch/site/assets/uploads/2018/05/SYR_AR5_FINAL_full_wcover.pdf

Kuhn A, Heidecke C, Roth A et al (2010) Importance of resource management for livelihood security under climate change in southern Morocco. In: Speth P, Christoph M, Diekkrüger B (eds) Impacts of global change on the hydrological cycle in west and Northwest Africa. Springer, Heidelberg, pp 566–591

Lazarev G (2008) L'élevage pastoral dans les Hauts Plateaux de l'Oriental du Maroc. Les notes d'analyse du CIHEAM 37:1–32

Maâtougui A, Acherkouk M, El Fadili M, Elhoumaizi MA (2011) Les pâturages steppiques de l'Oriental marocain. L'essentiel sur l'état de dégradation et les voies d'amélioration. Division de l'information et de la communication, INRA-Edition, Rabat, pp 1–57

Mabe FN, Sienso G, Donkoh S (2014) Determinants of choice of climate change adaptation strategies in Northern Ghana. Res Appl Econ 6(4):75–94

Mahdi M (2007) Pastoralism and institutional change in the oriental. In: Gertel J, Breuer I (eds) Pastoral Morocco: globalizing scapes of mobility and insecurity. Reichert, Wiesbaden, pp 93–105

Mahyou H, Tychon B, Balaghi R, Mimouni J, Paul R (2010) Désertification des parcours arides au Maroc. Tropicultura 28(2):107–114

Mahyou H, Tychon B, Balaghi R, Louhaichi M, Mimouni J (2016) A knowledge-based approach for mapping land degradation in the arid rangelands of North Africa. Land Degrad Dev 27:1574–1585

McNamara KT, Wetzstein ME, Douce GK (1991) Factors affecting peanut producer adoption of integrated pest management. Rev Agric Econ 13:129–139

Melhaoui M, Mezrhab A, Mimouni J (2018) Evaluation et cartographie de la sécheresse météorologique dans les hauts plateaux de l'Oriental du Maroc. Rev Microbiol Ind San Environ 12(1):71–92

Mokssit A (2012) Le point sur le changement climatique au Maroc. In: Ellinor ZM, Bennis A (eds) Environnement et changement climatique au Maroc- Diagnostic et Perspectives-. Konrad-Adenauer-Stiftung e.V., Bureau au Maroc, Rabat, pp 35–40

Moroccan Meteorological Office (2007) Les changements climatiques au Maroc: Observations et projections. DMN (Direction de la Météorologie nationale). Météo Maroc, Casablanca. https://www.preventionweb.net/files/changements_climatiques-214.pdf

Morocco (2011) Projet d'intégration du changement climatique dans la mise en œuvre du plan Maroc vert (Project to integrate climate change into the implementation of the Green Morocco Plan). Ministry of Agriculture and Maritime Fisheries and the Agency for Agricultural Development. http://www.ada.gov.ma/uplds/pars/ECIESPICCPMV.pdf

Morocco (2016) 3ème Communication Nationale du Maroc à la Convention Cadre des Nations Unies sur les changements climatiques (The third national communication of Morocco to the United Nations Framework Convention on Climate Change). Ministry Delegate to the Minister of Energy, Mines, Water and Environment in charge of the Environment. https://unfccc.int/resource/docs/natc/marnc3.pdf

Nabikolo D, Bashaasha B, Mangheni MN, Majaliwa JGM (2012) Determinants of climate change adaptation among male and female headed farm households in eastern Uganda. Afr Crop Sci J 20(s2):203–212

National Intelligence Council (NIC) (2009) North Africa: the impact of CC to 2030 (Selected Countries). A Commissioned Research Report NIC 2009-007D. https://www.dni.gov/files/documents/climate2030_north_africa.pdf

Ndamani F, Watanabe T (2016) Determinants of farmers' adaptation to climate change: a micro level analysis in Ghana. Sci Agric 73(3):201–208

Obayelu OA, Adepoju AO, Idowu T (2014) Factors influencing farmers' choices of adaptation to climate change in Ekiti State, Nigeria. J Agric Environ Int Dev 108(1):3–16. https://doi.org/10.12895/jaeid.20141.140

Opiyo F, Wasonga OV, Nyangito MM, Mureithi SM, Obando J, Munang R (2015) Determinants of perceptions of climate change and adaptation among Turkana pastoralists in northwestern Kenya. Clim Dev. https://doi.org/10.1080/17565529.2015.1034231

Ostertagova E, Ostertag O, Kováč J (2014) Methodology and application of the Kruskal-Wallis test. Appl Mech Mater 611:115–120

Ouédraogo M, Dembélé Y, Somé L (2010) Perceptions et stratégies d'adaptation aux changements des précipitations: cas des paysans du Burkina Faso. Sécheresse 21(2):87–96. https://doi.org/10.1684/sec.2010.0244

Paeth H, Born K, Girmes R et al (2009) Regional climate change in Tropical and Northern Africa due to greenhouse forcing and land use changes. J Clim 22:114–132

Piya L, Maharjan KL, Joshi NP (2013) Determinants of adaptation practices to climate change by Chepang households in the rural Mid-Hills of Nepal. Reg Environ Chang 13:437–447. https://doi.org/10.1007/s10113-012-0359-5

Rachik H (2007) Nomads: but how? In: Gertel J, Breuer I (eds) Pastoral Morocco – globalizing scapes of mobility and insecurity. Reichert, Wiesbaden, pp 211–252

Schilling J, Freier KP, Hertig E, Scheffran J (2012) Climate change, vulnerability and adaptation in North Africa with focus on Morocco. Agric Ecosyst Environ 156:12–26

Taruvinga A, Visser M, Zhou L (2016) Determinants of rural farmers' adoption of climate change adaptation strategies: evidence from the Amathole District Municipality, Eastern Cape Province, South Africa. Int J Environ Sci Dev 7(9):687–692. https://doi.org/10.18178/ijesd.2016.7.9.863

Tesfahunegn GB, Mekonen K, Tekle A (2016) Farmers' perception on causes, indicators and determinants of climate change in northern Ethiopia: implication for developing adaptation strategies. Appl Geogr 73:1–12

Tiwari KR, Rayamajhi S, Pokharel RK, Balla MK (2014) Determinants of the climate change adaptation in rural farming in Nepal Himalaya. Int J Multidiscip Curr Res 2:234–240

Yila JO, Resurreccion BP (2013) Determinants of smallholder farmers' adaptation strategies to climate change in the semi-arid Nguru Local Government Area, Northeastern Nigeria. Manag Environ Q Int J 24(3):341–364

Yohe G, Malone E, Brenkert A et al (2006) Global distributions of vulnerability to climate change. Integr Assess 6:35–44

# Permissions

All chapters in this book were first published by Springer; hereby published with permission under the Creative Commons Attribution License or equivalent. Every chapter published in this book has been scrutinized by our experts. Their significance has been extensively debated. The topics covered herein carry significant findings which will fuel the growth of the discipline. They may even be implemented as practical applications or may be referred to as a beginning point for another development.

The contributors of this book come from diverse backgrounds, making this book a truly international effort. This book will bring forth new frontiers with its revolutionizing research information and detailed analysis of the nascent developments around the world.

We would like to thank all the contributing authors for lending their expertise to make the book truly unique. They have played a crucial role in the development of this book. Without their invaluable contributions this book wouldn't have been possible. They have made vital efforts to compile up to date information on the varied aspects of this subject to make this book a valuable addition to the collection of many professionals and students.

This book was conceptualized with the vision of imparting up-to-date information and advanced data in this field. To ensure the same, a matchless editorial board was set up. Every individual on the board went through rigorous rounds of assessment to prove their worth. After which they invested a large part of their time researching and compiling the most relevant data for our readers.

The editorial board has been involved in producing this book since its inception. They have spent rigorous hours researching and exploring the diverse topics which have resulted in the successful publishing of this book. They have passed on their knowledge of decades through this book. To expedite this challenging task, the publisher supported the team at every step. A small team of assistant editors was also appointed to further simplify the editing procedure and attain best results for the readers.

Apart from the editorial board, the designing team has also invested a significant amount of their time in understanding the subject and creating the most relevant covers. They scrutinized every image to scout for the most suitable representation of the subject and create an appropriate cover for the book.

The publishing team has been an ardent support to the editorial, designing and production team. Their endless efforts to recruit the best for this project, has resulted in the accomplishment of this book. They are a veteran in the field of academics and their pool of knowledge is as vast as their experience in printing. Their expertise and guidance has proved useful at every step. Their uncompromising quality standards have made this book an exceptional effort. Their encouragement from time to time has been an inspiration for everyone.

The publisher and the editorial board hope that this book will prove to be a valuable piece of knowledge for researchers, students, practitioners and scholars across the globe.

# List of Contributors

**Keith Phiri and Philani Moyo**
Fort Hare Institute of Social and Economic Research (FHISER), University of Fort Hare, East London, South Africa

**Sibonokuhle Ndlovu and Moreblessings Mpofu**
Department of Development Studies, Faculty of Humanities and Social Sciences, Lupane State University, Bulawayo, Zimbabwe

**Henri-Count Evans**
School of Applied Human Sciences, Centre for Communication, Media and Society, University of Kwazulu-Natal, Durban, South Africa

**Doris Akachukwu**
Department of Agriculture, Health and Environment, Natural Resources Institute, University of Greenwich, Chatham, UK

**Michael Adedapo Gbadegesin**
Department of Biochemistry, Faculty of Basic Medical Sciences, University of Ibadan, Ibadan, Oyo State, Nigeria

**Philippa Chinyere Ojimelukwe**
Department of Food Science and Technology, Michael Okpara University of Agriculture, Umudike, Abia State, Nigeria

**Christopher John Atkinson**
Natural Resources Institute, University of Greenwich, London, UK
Department of Agriculture, Health and Environment, Natural Resources Institute, University of Greenwich, Chatham, UK

**Jonty Rawlins**
Natural Resources, OneWorld Sustainable Investments, Cape Town, South Africa

**Felix Kanungwe Kalaba**
School of Natural Resources, Copperbelt University, Kitwe, Zambia

**Gamuchirai Mutezo and Jean Mulopo**
School of Chemical and Metallurgical Engineering, Faculty of Engineering and Built Environment, University of the Witwatersrand, Johannesburg, South Africa

**Dumisani Chirambo**
Seeds of Opportunity, Blantyre, Malawi

**C. V. Nnamani**
Plant Systematics and Conservation Biology Research Unit, Department of Applied Biology, Faculty of Science, Ebonyi State University, Abakaliki, Nigeria

**D. B. Adewale**
Department of Crop Science and Horticulture, Federal University Oye-Ekiti, Ikole-Ekiti, Nigeria

**H. O. Oselebe**
Department of Crop Production and Landscape Management, Ebonyi State University, Abakaliki, Nigeria

**C. J. Atkinson**
Natural Resources Institute, University of Greenwich, London, UK
Department of Agriculture, Health and Environment, Natural Resources Institute, University of Greenwich, Chatham, UK

**Rasheedat Alliagbor**
Department of Agricultural and Resource Economics, Federal University of Technology Akure, Akure, Ondo, Nigeria

**David Olufemi Awolala**
African Climate Change Leadership Program (AfriCLP), University of Nairobi, Kenya and Institute of Resource Assessment, University of Dar es salaam, Dar es Salaam, Tanzania

**Igbekele Amos Ajibefun**
Department of Agricultural Economics and Rural Sociology, Auburn University, Auburn, AL, USA

**Olga Nekesa Mulama and Caroline Wanjiru Kariuki**
Strathmore Institute of Mathematical Sciences, Nairobi, Kenya

**Idowu Ologeh**
Department of Environmental Management and Toxicology, Lead City University, Ibadan, Nigeria

**Francis Adesina**
Department of Geography, Obafemi Awolowo University, Ile-Ife, Nigeria

**Victor Sobanke**
Research and Planning Department, National Centre for Technology Management, South West Office, Lagos, Nigeria

**Olushola Fadairo, Olufolake Adelakun, Siji Olutegbe and Oluwaseun Adeleke**
Department of Agricultural Extension and Rural Development, University of Ibadan, Ibadan, Nigeria

**Samuel Olajuyigbe**
Department of Forest Production and Products, University of Ibadan, Ibadan, Nigeria

**Tolulope Osayomi**
Department of Geography, University of Ibadan, Ibadan, Nigeria

**Olanrewaju Olaniyan**
Department of Economics, University of Ibadan, Ibadan, Nigeria

**Jemimah Timothy Ekanem**
Department of Agricultural Economics and Extension, Faculty of Agriculture, Akwa Ibom State University, Uyo, Nigeria

**Idongesit Michael Umoh**
Agricultural Science Education Unit, Department of Science, Redemption Academy, Uyo, Nigeria

**M. O. Abioja and J. A. Abiona**
Department of Animal Physiology, College of Animal Science and Livestock Production, Federal University of Agriculture, Abeokuta, Nigeria

**Wadii Snaibi**
Laboratory Communication, Education, Digital Usage and Creativity, ETIGGE Research Team, Bd. Mohammed VI, University Complex, Mohammed Premier Oujda University, Oujda, Morocco
National Institute of the Agronomic Research of Morocco, CRRA of Oujda, Oujda, Morocco

**Abdelhamid Mezrhab**
Laboratory Communication, Education, Digital Usage and Creativity, ETIGGE Research Team, Bd. Mohammed VI, University Complex, Mohammed Premier Oujda University, Oujda, Morocco

# Index

**A**

Adaptive Capacity, 2, 5, 18, 28, 45, 53, 56, 105-106, 112, 133, 147, 213, 215, 220, 225

Agricultural Production, 19, 40, 67, 105-106, 130, 146, 149, 154, 165, 169, 171

Arginine, 93, 96

**B**

Biodiversity, 4, 51, 60, 88, 91, 103, 146, 169

Broiler Chickens, 190-193, 195-198, 200-211

**C**

Carbohydrate, 26, 201, 210

Carbon Dioxide, 26-29, 36, 82, 120, 129, 136

Cassava, 25-26, 39, 49, 91, 93, 96, 98, 102, 135-139, 145, 157

Climate Change, 1-12, 15-22, 24, 26-29, 36-62, 64-65, 67, 76-80, 82-91, 93, 96, 98-99, 102-106, 113-122, 124, 128-139, 178-179, 182, 185-188, 190-196, 198, 200-201, 206-210, 212-215, 220, 222-223, 225, 228-231

Climate Hazards, 44, 48, 60, 170-171, 173, 175

Climate Variability, 5, 45, 58, 89-90, 105, 130, 132, 142, 144, 147-148, 151-152, 154, 163-165, 167-169, 171, 173, 175-176, 180, 182, 212, 215-218, 220, 222-223, 227, 229

Crop Failure, 8, 50, 136-139

Crop Productivity, 160

Crop Rotation, 137-138

Crop Yields, 12, 137, 169

Cultivars, 91, 185

**D**

Deforestation, 49-51, 58-59, 120, 145-146, 149, 152, 159, 161, 163, 165, 192

Desert Encroachment, 136, 192

Desertification, 90, 145, 149

Droughts, 2, 6, 45, 49, 52, 54, 82, 106, 117, 145, 169, 171, 212-215, 228

Dry Season, 107, 139, 145-146, 156, 200, 202, 207-208, 210

Dry Spells, 5, 45

**E**

Ecological Zones, 156-157, 161, 163

Ecosystem, 4, 25, 28-29, 49, 59-60, 88, 144-147, 153, 158, 160, 164, 166, 169, 212, 214, 218

Egg Production, 191, 194, 198, 202, 207

Environmental Degradation, 4, 35-36, 55, 88, 145, 149, 153, 161-162

Environmental Health, 36

Environmental Temperature, 191, 193-194, 196, 198, 201

Extreme Temperatures, 5, 45, 116

**F**

Farming Practices, 51, 117

Farming Systems, 49, 105, 117, 165

Food Security, 2, 16, 28-29, 32, 35, 53, 56, 58, 87, 89-92, 96, 101-102, 108, 112-114, 119, 144, 146, 148-149, 152, 160, 165-166, 188, 210

**G**

Genetic Erosion, 88, 91, 99

Genetic Resources, 87-88, 91-92, 99, 101-102

Germplasm, 88, 91, 99-103

Global Temperature, 192, 194

Global Warming, 22, 24, 26, 29, 36, 38-39, 59-61, 88, 102, 117, 119, 130, 191-192, 207

Glycine, 93

Greenhouse Gas, 26, 28, 38, 59-60, 64, 68, 86, 88, 102, 132-133

Gross Domestic Product, 117

Groundnut, 93, 195

**H**

Heat Stress, 45, 190-191, 196-197, 208-210

Heavy Rainfall, 14, 45

Herbicides, 138, 163-164

Histidine, 93, 96

Horticulture, 87

Humidity, 107, 116, 194, 198, 207

## I
Indigenous Technology, 40, 131-134, 140-141
Intercropping, 137-138

## L
Land Degradation, 42, 51, 137, 146, 209, 230
Land Improvement, 132, 139

## M
Methane, 26, 63, 68-69, 82
Methionine, 93, 201
Minerals, 91, 93, 204

## N
Natural Disasters, 18, 64, 80, 99, 151, 181, 187
Neem Seed, 132, 136-137, 140
Nitrogen Fixation, 99, 102-103
Nitrous Oxide, 82
Nutrient Cycling, 146
Nutrition, 2, 19, 35, 40, 61, 88, 149, 200, 205

## O
Organic Agriculture, 136
Organic Matter, 69, 99, 136, 216
Oxygen, 28, 205

## P
Pest Invasion, 136-137
Pest Management, 230
Pesticide, 137
Pesticides, 138-139
Phosphorus, 39
Precipitation, 14-15, 27, 124, 154, 156
Protein, 91, 93, 100, 191-192, 200-201, 209
Proteins, 26, 100
Public Health, 36, 40, 53, 58, 64, 118, 130, 166, 205

## R
Rainwater Harvesting, 19, 37, 132, 136
Rainy Season, 107, 109, 195
Rationale, 56, 65, 76, 80, 121
Renewable Energy, 40, 66, 169
Rural Population, 2, 106, 167-168, 171

## S
Sea Level, 4, 24, 183
Semiarid Regions, 6, 195
Small-scale Farmers, 128
Smallholder Farmers, 2, 5-9, 11, 15-16, 18-19, 56, 106, 131, 134-140, 142, 208, 231
Soil Erosion, 99, 132, 136, 138-140
Soil Infertility, 136-137, 162
Soil Water, 29, 50, 99
Sphenostylis Stenocarpa, 88, 92-94, 101-103
Survivability, 191, 204
Sustainability, 3, 5, 59, 66, 80, 83-86, 102, 113, 142, 212, 214-215, 229
Sustainable Agriculture, 18, 146-147, 149
Sustainable Development, 19, 40, 43-44, 46, 54-59, 65, 67, 85, 88, 102, 114, 116, 132, 143, 145, 147, 149, 164, 207

## T
Thermal Conditioning, 191, 203-204, 211

## W
Water Catchment, 122
Water Resources, 121, 216
Weather Patterns, 6, 117-119, 121, 128-129, 160, 169
Wet Season, 47-48, 156, 208

## Y
Yam Bean, 87-88, 92-97, 99-103

www.ingramcontent.com/pod-product-compliance
Lightning Source LLC
LaVergne TN
LVHW080622210926
848485LV00003B/194